CAPITAL INVESTMENT MODELS
OF
THE OIL AND GAS INDUSTRY

This is a volume in the Arno Press collection

ENERGY
IN
THE AMERICAN ECONOMY

Advisory Editor
Stuart Bruchey

Research Associate
Eleanor Bruchey

See last pages of this volume for a complete list of titles

CAPITAL INVESTMENT MODELS
OF
THE OIL AND GAS INDUSTRY

A Systems Approach

Louis John Joseph Allain

ARNO PRESS
A New York Times Company
New York • 1979

Editorial Supervision: DIETRICH SNELL

First published in book form 1979 by Arno Press Inc.

ENERGY IN THE AMERICAN ECONOMY
ISBN for complete set: 0-405-11957-7
See last pages of this volume for titles.

Manufactured in the United States of America

Library of Congress Cataloging in Publication Data

Allain, Louis John Joseph.
    Capital investment models of the oil and gas industry.

    (Energy in the American economy)
    Originally presented as the author's thesis, Purdue
University, 1970.
    Bibliography: p.
    1. Petroleum industry and trade--United States--
Finance. 2. Gas industry--United States--Finance.
3. Oil fields--Production methods--Costs. 4. Capital
investments--Mathematical models. I. Title. II. Se-
ries.
HD9564.A78 1979   338.2'3   78-22654
ISBN 0-405-11959-3

CAPITAL INVESTMENT MODELS OF THE

OIL AND GAS INDUSTRY: A SYSTEMS APPROACH

A Thesis

Submitted to the Faculty

of

Purdue University

by

Louis John Joseph Allain

In Partial Fulfillment of the

Requirements for the Degree

of

Doctor of Philosophy

January 1970

ACKNOWLEDGMENTS

To my major professor, Keith Brown, I am especially indebted for first his suggestion of the topic, and secondly for his encouragement, interest, and guidance throughout the writing of this dissertation.

To Professors Claude Colantoni, Mohammed El Horidi, and Floyd Gillis I am indebted for their guidance and assistance in the completion of this document, and for serving on my thesis committee.

The assistance of the typist, Geraldine McGee, for typing the manuscript is deeply appreciated in view of the long arduous task which faced her.

Finally to my family, I am most deeply indebted to my mother and father for their encouragement, guidance, and assistance throughout my period of study. Likewise, my brother served as a source of encouragement and counselor during my graduate studies and for his interest and guidance I am deeply grateful. Lastly, my sisters were a continual source of joy in our many conversations.

TABLE OF CONTENTS

LIST OF TABLES

LIST OF FIGURES

ABSTRACT

Allain, Louis John Joseph. Ph.D., Purdue University, January 1970. Capital Investment Models of the Oil and Gas Industry: A Systems Approach. Major Professor: Keith C. Brown.

The problem considered in this thesis is that of applying modern system's theory to the problem of investment, production, and capital budgeting in the oil and gas industry.

Historically, the problem has its origins in the engineering problem of well spacing. At first, the problem was considered to be solely a physical problem of placing enough wells on a reservoir to achieve the optimum physical ultimate recovery. However, because of belief in Cutler's Rule optimum physical recovery could only be achieved by close spacing. Once the basic physical considerations underlying this theory was proved false, more consideration was then given to the economic problems of economic ultimate recovery and costs.

Before considering the economics of the problem, we discuss the present day state of the art of petroleum engineering in regard to the calculation of oil and gas reserves and the principles of fluid flow. This is necessary for in order to determine the economics of ultimate recovery, it is important to determine the principles used in calculating reserves, and understanding the process of fluid flow through a porous media.

The first economic considerations are presented in a diagramatic mode. The problem is defined in terms of three inter-related problems, the rate of production, the economic ultimate recovery, and the economics of well spacing. Having determined the maximum efficient rate of production, optimal well spacing is then determined after consideration of revenues and costs.

From simple diagramatic models of efficient production, we expand the analysis to consider linear and dynamic models of production and extraction from a reservoir. We begin with a linear one well, one reservoir presentation and advance the exposition to include multi well, multi reservoir situations, allowing also for well interference. We close the presentation of production models by considering a dynamic model which yields the optimal timing and phasing of wells under the dynamic assumptions and restrictions of the model.

Our next concern is with the presentation of investment and budgeting models for the oil and gas industry. The models begin with simple linear models for investment, and advance to complex dynamic models which yield the optimal timing and phasing of investment projects over the life of the reservoir.

The method of solution for these models is programming techniques. For the simple linear models, the Simplex Algorithum or some modification of this technique is necessary to generate optimal solutions. To solve the dynamic problems it is necessary to use the principle of optimality. A branch-bound algorithurm using this principle is developed and programmed to obtain optimal solutions to the dynamic models as formulated.

Finally, in light of the models developed, conclusions are reached concerning the use of programming techniques in economics. Secondly, policy recommendations are made in regard to pricing, taxation, optimal investment, and depletion allowance in the oil and gas industry. Thirdly, extensions to the analysis are given and recommendations for future research in this area made in light of the analysis presented.

CHAPTER 1

INTRODUCTION

Exposition of the Problem and Methodology

There exists in economics and the applied management fields much concern with, and use of mathematics both
applied and pure, to attempt to understand business problems both in the macro and micro sense. The type of
mathematics used by the modern industrial analyst ranges
over the entire array of pure and applied mathematics.
This thesis attempts to use one of the new types of tools
recently developed, namely, it attempts to apply linear and
dynamic programming to economic analysis.

Much theoretical work has been completed on the
pure mathematical aspects of programming techniques, and in
recent time more concern is being paid to applying these
techniques to real problems, and particularly to the area
of management science. This work is not a treatise on the
theoretical aspects of programming or management science,
but rather recognizes that over time many of the theories

of management science have to be applied to real industrial problems to show that the theory is relevant. This then is the concern of this thesis, the application of some theoretical techniques to applied industrial areas and problems.

This thesis is an application of one view of philosophy that is prevelant in management science today, namely that school which holds that if theory is to be useful or relevant in that science it must be applied to a real class of problems. However, it differs from this school somewhat in that it approaches the systems problem from a more diversified point of view. Namely, at the initial stages of the problem, historical research surrounding the analysis is considered important in providing the setting for the future analysis.

Once the historical framework of the problem has been defined and analyzed, then the analysis takes the more traditional approach to solving the problem. That is, a review of the relevant theoretical and applied work on the problem is undertaken to understand the relevant theory, and to analyze the contributions and the failures of these contributions in the context of solving the problem at hand. Next, applications and/or extensions of the theory are undertaken to show first the value of the theory

and/or to show that the new theory is better than the old
and applies to a more general class of problems.

We have attempted to follow here this general
philosophy. Basically, we are interested in the application of programming techniques to economic and management
problems. The concern of the thesis will not be the
extension of the techniques but rather the application of
these same techniques to a class of problems. The problems
chosen were those of the oil and gas industry, individually
the most important industry in this country. The particular
area of concern was a study of the productive stage of the
industrial process and specifically the investment and well
spacing problem in this domain.

There are four stages in the gas-oil production
process; production, or extraction from the subsurface:
refining: marketing: and transportation. Much work has been
completed on the refining, marketing and transportation of
the crude. For examples of the work published on these
areas, the reader is referred to the references cited
below.[1]

---

[1]W. Garvin, H. Crandall, et. al., "Application of
Linear Programming in the Oil Industry," Management
Science, (July, 1957), Vol. 3.

However, little work has been undertaken or completed on the economic problems encountered in the production/extraction stage. Thus, while much work has been done on capital theory, investment, and budgeting problems, these important tools have not been applied to the area of production/extraction in the oil-gas industry.

The necessity of considering applications in this area is obvious and important. As we shall see later, one

---

A. Charnes, W. W. Cooper and B. Mellon, "Blending Aviation Gasolines," Econometrica, (April, 1952), Vol. 20.

F. L. Hitchcock, "The Distribution of a Product from Several Sources to Numerous Localities," Journal of Math and Physics, (1941), Vol. 20.

G. B. Dantzig and D. R. Fulkerson, "Minimizing the Number of Tankers to Meet a Fixed Schedule," Naval Research Logistics Quarterly, (1954), Vol. 1.

G. B. Dantzig and J. H. Ramser, "The Truck Dispatching Problem," Management Science, (1959), Vol. 6.

A. S. Manne, Scheduling of Petroleum Refinery Operations, (Cambridge, Mass., Harvard University Press, 1956).

J. S. Aronofsky and A. C. Williams, "The Use of Linear Programming and Mathematical Models in Underground Oil Production, Management Science, (1962), Vol. 8.

R. A. Catchpole, "The Application of Linear Programming to Integrated Supply Problems in the Oil Industry," Operational Research Quarterly, (1962), Vol. 13.

Plus other references not cited here, but contained in the main body of the thesis.

half of the total amount of money spent by the oil-gas
industry takes place in the production/extraction stage in
the form of expenditures on wells and productive facilities.
Gross inefficiencies are known to exist in this area. The
industry itself and its principal organs, the API and AGA,
have conducted extensive studies on well-spacing and investment in production facilities. Likewise, interesting
ramifications in terms of policy measures arise when considering the production/extraction stage in the form of
recommendations concerning depreciation allowances and
depletion allowances granted at this stage of production to
the industry.

Because of the importance of the problem and the
concern and advantages which would result from association
and cooperation with gas and oil producers, the study was
intended to be undertaken in collaboration with producers
in the industry. However, at this point in time and
maturity of the industry, little interest was shown in this
study, and no cooperation was obtainable in procuring
relevant data which would have added realism to the study.
Thus, it was necessary to simulate the models herein
derived. The dynamic models were simulated primarily
because the more interesting case in economics is the

dynamic case, and since time and time discounting were so integral to the analysis of the problem, it was considered that these models would be more important and more realistic to simulate. It is urged that as a follow-up to this presentation that the industry become concerned with the problems herein considered and that variations of the techniques employed in this thesis, if not already con-sidered by their research departments, be undertaken to further the understanding and techniques of analysis which can and should be applied to the production/extraction stage.

Organization

The thesis can be loosely divided into four independent sections. The sections can be tentatively designated for expository purposes as the historical pre-sentation, the physical principles of production section, the relevant economic and management principles, theories, and models exposition, and lastly the application and computational simulations of the model.

In Chapter 2 we begin with the historical presentation of the investment-budgeting problem. The origin of the problem in the gas-oil industry is found in well-spacing and facilities necessary for extraction or production. The initial treatment of the problem was basically concerned only with the physical-engineering aspects of the problem. In the early literature the treatment was basically dominated by Cutler's rules and the Jarmin effect. Until the basic physical considerations underlying these two theories were proved false, little consideration was given to the economic aspects of the problem. With these two theories disposed of by advances in the mathematical theory of fluid flow more consideration was given to the economic problems existing in this stage. In the physical and engineering literature the best statement of the problem and statement of the economics principles was given by Muscat in 1949.

The 1949-50 period then serves as the breaking point in the industry where research turned toward consideration of economic recovery and away from the idea of linking well-spacing with physical ultimate recovery. The engineering literature concentrated on the physical principles of recovery and the determination of characteristics and principles of flow mechanics. The problem of well-spacing was left then to the analysis of the economist.

In Chapters 3 and 4 then we discuss the present day state of the art concerning the physical aspects of the

gas-oil productive process. Both gas and oil are discussed
since once either the gas or oil process is understood the
other naturally follows. The section is divided into the
physical aspects and principles of fluid recovery, and the
flow of fluid characteristics. This exposition of these
physical and engineering processes is deemed necessary
since in determining the economics of ultimate recovery it
is necessary to understand the concepts and principles used
in calculating reserves, and likewise the flow of a fluid
or gas through a porous media.

Though the physical principles of oil recovery are
necessary to the study of the spacing problems, the important determinants of the problem are the economic factors.
In Chapter 5 then we first present in a diagramatic mode
the relevant economic concepts and considerations of the
problem. The problem is defined in terms of three interrelated problems, the rate of production, the economic
ultimate recovery, and the economics of well-spacing. It
is necessary in this context to define then the maximum
effectent rate of production. Once this is determined, it
is possible to define the ultimate economic recovery and
the well spacing for a reservoir in terms of diagrams.

Having determined the MER, it is necessary then to discuss and introduce the various forms which the revenue function and cost functions can take, for these will be used continuously in the model presentation and an understanding of these forms and functions is vital to understanding the various forms of the objective functions can take depending on the assumptions and needs of the model. If the diagramatic method of analysis shows us anything, it is the limitedness of the graphical method of solution to the problem we are considering.

Thus, in Chapter 6 we begin by considering some simple models surrounding the production/extraction of gas and oil from a reservoir(s). We begin with the simple presentation of linear models, one reservoir, one well, and advance our exposition to include multi-reservoir, multi-well situations, allowing in the analysis for well interference. We close this initial presentation of production models by consideration of a simple dynamic model which yields in the optimal timing and phasing of wells under the assumptions, presentations, and restrictions of the model given in that chapter.

Before considering investment and budgeting models, in Chapter 7 we present a brief synopsis of the relevant literature in investment and capital theory as well as some major studies undertaken in capital budgeting of concern in this study. The review covers investment and capital theory from Marshall-Fisher to V. L. Smith and his most recent contribution to natural resources. Likewise budgeting is covered from Dean-Markowitz to the most recent works by Naslund and Techiroew and their most recent contributions to the field of capital budgeting. The rationale of this section and its being placed between the models of production and investment is that this serves as the foundation for Chapters 8 and 9 and thus, the ideas used in these chapters depend very heavily on the theoretical works of the men reviewed.

Chapter 8 again continues the presentation of models of investment and budgeting for the oil and gas industry. The analysis is the same as before, beginning with simple linear models and advancing to the more complex dynamic models at the end of the chapter. The final section of the chapter presents a model for timing and phasing of investment projects over the life of the reservoir for the firm.

After discussing the theoretical and mathematical aspects of the model, we enter the final stage of the

thesis, the computational presentation. In order to solve
the problem it is necessary to have a method of solution
or solutions--algorithm(s). For the linear models, the
simple algorithm, the revised simplex, and parametric
techniques are necessary to generate optimal solutions.
These techniques and algorithms are not presented since
they are readily available in most programming text. To
solve the dynamic problem as formulated, Chapter 9 considers
and uses Bellman's principles of optimality and modifies
the Bellman algorithm with the necessary modifications to
present solutions to the problem as formulated. The solution takes the form of a branch-bound algorithm. Detailed
analysis is given of the modifications and the relevant
definitions and theorems necessary to ensure that
optimality is always maintained throughout. A computer
program is given in detail and is available upon written
request. Solutions obtained are presented in Appendices
B-J.

The conclusions along with policy recommendations
are formally presented in Chapter 10. It is important to
realize that the thesis has been concentrating on the presentation of techniques and applying these techniques to
economic analysis. The main conclusion of this thesis can

then be best put by saying that dynamic programming is a useful and helpful technique for the economist and deserves more and careful consideration than the profession presently has given it.

CHAPTER 2

WELL SPACING IN THE ENGINEERING LITERATURE

Introduction

The engineering and physical problems surrounding well spacing, the rate of production, and ultimate recovery from a reservoir have plagued the petroleum industry and conservationalists for some time. Prior to the initiation of state regulations in the 1930's, almost unregulated production practices prevailed in this country, and the general rule followed by the industry was the rule of capture, seizure of all the product possible regardless of time by drilling the reservoir most extensively. Thus, with the discovery of new pools, the method of development was to drill as extensively as possible each individual pool, and thus close well spacing became a commonly accepted trade practice. The general purpose of this extensive drilling was to obtain as much product as possible from under each operator's land and to thus prevent the drainage of oil to neighboring plots. This lead to the practice of drilling

offset wells that is, wells drilled in the same pool such
that they compensate each other in terms of recoverable
reserves, and to the development of each tract of land to at
least the same stage of development as prevailed on adjacent
leases. Accompanying this extensive drilling was production of each well at maximum capacity, which many times
without the money for proper storage facilities meant
literally streams of waste, in the form of crude and escaping gas, of the valuable nonrenewable resource. Waste then
was the common word both above ground and underground in the
race to capture the available oil or gas.

Before 1920, many oil fields were drilled to
densities ranging from one well to five acres to one well to
every one or two acres. Examples of fields of these densities are: Spindletop, Sour Lake, Humble, and Goose Creek,
all in Texas. Some old spacings can be seen from Table 1.

As early as 1916, McMurray and Lewis, in a
technical publication of the Bureau of Mines, demonstrated
that many problems existed in the oil and gas industry, but
one of the main problems was over drilling, and in the
paper came up with a solution which has since then been
termed unitization of reserves, that is, operating the

15

carried over into the peace time and became reorganized as
the American Petroleum Institute, and since then has been
the representative body of the petroleum industry.

One of the directors of the Institute, Henry L.
Doherty, was a conservationist and was deeply concerned with
the problem of waste, and by himself had worked out his solution to the problem, which in many respects was similar to
the McMurray-Lewis solution, that is, Doherty called for
the unit-pool operation of oil and gas reservoirs under
federal jurisdication. While many at the API agreed with
Doherty, they were uniformly opposed to his solution in that
none wanted the federal government to become involved in the
industry.

Unable to gain support with the industry or with
the API, Doherty addressed a long letter to President
Coolidge in 1924. In this letter, he outlined in complete
detail the conditions as they existed in the industry.
Relying on what now has come to be known as "traditional"
arguments, that is, national security, conservation, and
costs of production, he argued that the petroleum producing
states be called upon to pass conservation laws, and that
if they failed to comply with the legislation, the federal
government should itself take over the regulation of the

reservoir as a single unit or body.[1] Continuing, they

recommended that this unitization of reserves be compulsory,

since at that time they realized that both the industry

giants and individual producers would oppose any control.[2]

Table 1 - Well Spacing In Old Fields

Field	Total Acres of Reservoir	Number of Wells	Well Spacing acre/well
Smackover, Ark.	29,500	3,919	7.5
El Dorado, Ark.	10,650	1,125	9.5
Cushing, Okla.	27,800	3,731	7.5
Heallton, Okla.	7,200	2,511	2.9
Haynesville, La.	13,650	978	14.0
Homer, La.	3,020	651	4.6
Hendricks, Texas	9,800	621	15.8
Easton, Texas	650	1,136	.6
Spindletop, Texas	500	1,416	.3
Mexia, Texas	3,920	600	6.5

However, the advent of the war, and the mobiliza-

tion of the petroleum industry to the aid of the war effort,

prevented any stir of reaction to the McMurray-Lewis paper,

or any cognizance to the waste that existed in the industry.

The organization set up for the war purpose, however,

---

[1]W. F. McMurray and J. O. Lewis, "Underground Wastes in Oil and Gas Fields and Methods of Prevention," U. S. Bureau of Mines, Technical Paper, 1916.

[2]E. W. Zimmerman, Conservation In The Production of Petroleum, (New Haven, Yale University Press, 1957), p. 289.

industry.[3] Almost as a direct response to Doherty's request,
on December 19, 1924, Coolidge created the Federal Oil Conservation Board which consisted of the Secretaries of War,
Navy, Interior and Commerce. The Board's functions were to
oversee for the federal government the total conservation
problems in the petroleum industry.

By this Act, Coolidge ushered in the period in
which wasteful petroleum production practices were brought
to the fore, and as a result, a serious study of petroleum
production practices was finally initiated. This lead to
the period 1930-1931, when state regulation was seriously
debated and finally initiated in the various petroleum producing states.

Well Spacing Concepts: Engineering

The year 1924 also marks the date for the first
important study in well spacing, one which was to dominate
the industry for about two decades and which still today
finds proponents. The work was by W. W. Cutler, and his
maxim has become known as "Cutler's Rule." Cutler's
studies were done for the U. S. Bureau of Mines, and in

---

[3]For a copy of the original letter, one is referred
to: Robert E. Hardwicke, Antitrust Laws et. al. v. Unit
Operation of Oil or Gas Pools, American Institute of Mining
and Metallurgical Engineers (New York, 1948), pp. 179-190.

1924 when petroleum engineering was still in its infancy.
Cutler analyzed and compared production statistics for
several solutions -- gas drive pools with varying well
densities. Based on these observations, and without benefit
of an extensive knowledge of flow of fluids through porous
media, he proposed the following rule, hence termed
"Cutler's Rule:" "The ultimate productions for wells of
equal size in the same pool where there is interference
(as shown by a difference in the production decline curves
for different spacing) seem approximately to vary directly
as the square roots of the areas drained by the wells."[4]
Further in the same work, he restates his "rule" in the
following words: "The recovery from wells of equal size
producing under similar conditions in the same pool is
proportional to the average distance that the oil moves to
get to the well ... From this rule, it follows that where
interference exists between wells, doubling the distance
between wells doubles the ultimate production per well and
halves the ultimate recovery per acre."[5] The implication

_____

[4]W. W. Cutler, Jr., "Estimation of Underground Oil
Reserves by Oil-Well Production Curves," United States
Bureau of Mines, Bulletin No. 228. (Washington, D. C.,
1924), p. 89.

[5]Ibid., p. 89

of his "tentative rule" would be that a well with drainage radius of 330 ft. would ultimately produce one-half as much oil as a well with drainage radius of 660 ft.; and a 40 acre tract of land would produce two times as much oil from four wells as one well bore in the center of each 10 acre fraction, as it would from one well in the center of the 40 acres. Cutler attempted to justify his rule on the basis of his concept of energy relations, wherein he stated that the energy required to move a fluid through a pipe is proportional to the distance. He has since been shown to be erroneous in his interpretation of fluid movement, and the only thing he did apparently show was that drainage was possible over long distances if well interference existed, a conclusion opposite to what he proposed.

Cutler's rule has been widely misinterpreted as a rule to show that the greater the well density in a field the greater will be the ultimate oil recovery from that field. And there is clear implication in Cutler's Rule that close well spacing would lead to a larger ultimate pool recovery than wide spacing. It is important to realize that Cutler's Rule is really the rule of capture since his observations and statistics were based on fields with open-flow operations and at a time when maximum capture was the all important object of production.

However, as a result of Cutler's work, the industry was left with the impression that the number of wells drilled affects the efficiency with which reservoir energy is used to produce oil. Thus, the impression was left to the industry that the physical oil recovery from a reservoir was directly dependent on well spacing. The important point is that Cutler's work did become the center point for over thirty years for the proponents of close well spacing.

In another important work, S. C. Herold developed an analytical method of drainage for natural oil reservoirs. Herold divides producing wells into three clases, namely those producing from (1) hydraulic, (2) volumetric, and (3) capillary control. In his theoretical analysis, a well of either of the first two classes properly located in a field will eventually drain the entire reservoir. In these two cases then no well spacing needs to be calculated. In the capillary control wells, each well drains a definite area, and thus a well spacing rule is necessary to achieve

---

[6]S. C. Herold, Analytical Principles of the Production of Oil, Gas, and Water from Wells, (Palo Alto, Stanford University Press, 1928).

the greatest ultimate recovery. This method of well
drainage in these capillary wells has been termed "The
Janim Effect," or "the bubble effect" in the flow of liquid
and gas in a restricted capillary tube. The idea is that
the drainage area of a well is fixed within definite limits
by the resistance to flow caused by an accumulation of gas
bubbles within the oil sand. He further reasoned that at
abandonment there existed a point in the radius of drainage
beyond the well bore where the pressure balanced this bubble
resistance to flow and beyond which point no further depletion could occur.

In response to "Cutler's Rule" and Herold's
"Janim Action" there developed in the engineering literature
a concern with the mathematical derivation of a formula
which would determine well spacing within the limits of
these rules. One of the first papers was written by Robert
Phelps in which he attempted to develop formulas to permit
the computation of well spacing on oil reservoirs that
would result in the maximum net return per acre based on
both Cutler's and Herold's analysis.[7] The formula he

---

[7]Robert W. Phelps, "Analytical Principles of the
Spacing of Oil and Gas Wells" Trans AIME, 1928-1929, Vol.
82, p. 90-102.

arrives at based on Cutler's study is given by:

$$1 \qquad R_\$ = R_m = \frac{(A + Xa)^3}{Xa^2} - \frac{2VpavK}{CW}$$

where he defines:

$R_\$$ = net return per acre;

R_m = maximum return per acre;

Xa = well spacing in terms of acres per well;

$Vpav$ = present value of the available production per acre including pumping costs;

Cw = cost of drilling each well;

A, K = constants.

The formula he derives for proper well spacing based on Herold's theory of capillary control is given by equation:

$$2 \qquad S = 63.5 \sqrt[3]{\frac{RdCw}{Pav}}$$

where S is defined by Phelps as the well spacing, or the distance between two wells. In the remainder of the paper, Phelps is concerned with showing how his formulae actually proves that close well spacing actually enhances ultimate production.

In a series of papers, W. P. Haseman developed a formula for determining the proper well spacing and the rate of production for a reservoir. Following Cutler's

Rule and likewise Cutler's energy principles, Haseman states: "This principle governs the state and the change of state of fluids or fluid scriptures confined under pressure within a porous and permeable reservoir in such a way as to direct a flow from a reservoir into and out of a drilled well with a reduction of potential energy."[8]

His factors which affect well spacing are listed in two groups, those that are fixed and those variable. The fixed factors are defined as those which comprise the physical characteristics of the reservoir. These then can be assumed not to affect well spacing. The variable factors on the other hand are controllable at will, and well spacing is one of the most important determinants of these variable factors. Utilizing this distinction, he then arrives at his formula for well spacing which is given by

3
$$P_n = P_o (1 - e^{-\lambda n})$$

---

[8]W. P. Haseman, "Profits and Proper Spacing of Wells," Oil and Gas Journal, Oct. 18, 1928, Vol. 27, no. 22, p. 53.

_____, "A Formula Method for Well Spacing and Rate of Production," National Petroleum News, Vol. 21, no. 19, p. 59.

_____, "A Theory of Well Spacing," Trans. AIME, (1930), Vol. 86, p. 146-149.

where he defines:

>Po = total quantity of active oil under a
>given tract of land (active oil being
>defined as recoverable oil).

>Pn = quantity of active oil that is yielded
>by n wells spaced to the given tract of
>land.

>n = number of wells drilled.

Applying his formula to past reservoir data, Haseman modestly states: "This formula is the most useful in the science of oil production." According to him, the application of his method will yield an optimal program of development and the operation of the reservoir at maximum possible efficiency.

It is important to note the trends that exist in these first studies. First, the authors have reviewed, as well as many others, attempted to derive a single simple mathematical formula for well spacing. Secondly, they were concerned with the physical operation of the reservoir and the physical development and operation of a physical development program. However, the question of efficiency was being asked, and studies were being conducted by using data from existing developed fields.

In the early 1930's, Moore, Schilthuis, Hurst, and Muskat approached the problem of well spacing from a

new point of view. They offered mathematical analysis of
the pressure distribution, and the radius of influence
around a well during single-phase flow in the unsteady
state.[9] The results of their studies showed that the
pressure reduction starts with the well bore and progress-
ively moves outward to greater and greater distances from
the well bore. It appeared to these authors that there was
no limit to the range of ultimate pressure reduction and
reaction in a porous rock having continuous fluid trans-
missibility. Thus, they established that drainage is
unlimited in a continuous porous medium, and that well
spacing on a much wider basis than previously had been
postulated could adequately drain a field.

In 1937, Muskat capped his work on flow of fluids

---

[9]T. V. Moore, R. J. Schilthuis, and W. Hurst,
"The Determination of Permeability from Field Data,"
API Production Div. Proc. Bul., (1933), Vol. 211, p. 4.

W. Hurst, "Unsteady Flow of Fluids in Oil
Reservoirs," Physics, (1934), Vol. 5, p. 20.

M. Muskat, "The Flow of Compressible Fluids
Through Porous Media and Some Problems in Conduction,
Encroachment of Water Into an Oil Sand," Physics, (1934),
Vol. 5, p. 71.

M. Muskat and M. W. Meres, "The Flow of
Heterogeneous Fluids Through Porous Media," Physics, (1936),
Vol. 7, p. 325.

by publishing his now famous work on applied physics of
fluid flow.[10] Likewise in his 1940 article, he was still
only concerned with the physical properties and principles
involved in well spacing.[11] While Muskat contributed much
in the understanding of fluid flow through porous media in
unsteady state, in linear and radical flow, he viewed at
this time the problem primarily as a physical problem.
Thus, while these earlier works are important, they are too
technical to be matter of inclusion here.[12]

Some important new concepts were introduced into
well spacing literature by L. L. Foley in 1938.[13] Again
he reviewed Cutler's work and his efficient use of field
energy would come from utilizing water drive, and thus
the answer to obtaining efficiency was to utilize water
drive in every oil reservoir and to obtain the proper well

[10]M. Muskat, Flow of Homogeneous Fluids,
(New York, The MacMillan Co., 1937).

[11]_____, "Principles of Well Spacing,"
Trans. AIME, (1940), Vol. 136, p. 136.

[12]Muskat's 1949 review of petroleum technology
and the economic treatment he gives therein will be treated
in detail later.

[13]Lyndon L. Foley, "Spacing of Oil Wells,"
Trans. AIME, (1938), Vol. 127, p. 15-24.

spacing solution, it was necessary to divide the efficient rate of production for the field by the efficient rate for the average individual well. The resultant quotient reveals the number of wells that should be drilled. Further in the paper he insinuates that the spacing of wells should be such that the yield from the reservoir is in some relation to market demand, and that only enough wells should be drilled so that the wells could be utilized at their most efficient rate of production. However, toward the end of the paper he yields to the close drilling school when he says that the real answer to the spacing problem is: "To learn to drill more cheaply." Thus, if drilling becomes cheaper, it is possible to drill more wells and then Cutler's Rule will apply. Although he did ally himself with the "old close school," Foley did introduce the new concepts of efficient production in relation to market demand which later was the rule adapted by some state conservationist and commissions in the 1940's.

The most important work published in 1939 was that by Miller and Higgins.[14] In this report, Miller and Higgins

---

[14]H. C. Miller and R. V. Higgins, "Review of Cutler's Rule of Well Spacing and Use of Energy," Bureau of Mines Report of Investigation 3439, (1939).

acknowledged the error of Cutler's original assumption that
radical flow of fluid through a body of porous media obeyed
the same mechanical law with reference to energy required
as that for linear flow through a pipe. They showed that
for a liquid, or a dead oil, an equation based on Darcy's
Law would be applicable for a radical flow through uniform
sands. They worked out an example under an assumed hypothetical set of pressure and volume of oil production conditions and showed that only 10% more energy would be
required to move a given volume of oil in 24 hours to a
well with a drainage radius of 660 ft. than would be
required to move the same volume in 24 hours to a well with
drainage radius of 330 ft. According to Cutler's Rule,
doubling the radius flow would have required doubling the
energy.

Miller and Higgins also question the sufficiency
for purpose of general application of the factual oil production data cited by Cutler as a basis of statistical com-
parisons, that furnished a background for his original
assumption of a square root relationship between ultimate
production from differently spaced wells in the same field.
The effect of the Miller-Higgins paper is that it totally
annulled "Cutler's tentative Rule."

The Miller-Higgins paper caused more than a stir in the petroleum industry, although Muskat et. al. were indicating as much as far back as 1934. However, the API deemed Miller and Higgins' results revolutionary enough to appoint a blue ribbon committee to investigate the problem of well spacing. The first report by the committee was in 1939, the second in 1940. These being preliminary, it is of no concern to us to review them.

However, the 1942 report on operating engineering principles, as well as economic and legal practices of well spacing, is important since it considers for the first time the well spacing problem in its total dimensions.[15]

Since the report is quite long, it is best to summarize the work of the Committee by giving the conclusions arrived at in the study. The first conclusion of the Progress Report, was of course, that conservation is necessary, first, for the purpose of providing an adequate supply of product in the future at reasonable cost, and secondly, to protect the property rights of owners of tracts

---

[15]Progress Report on Standards of Allocation of Oil Production Within and Among Pools, The Special Study Committee and Legal Advisory Committee on Well Spacing and Allocation of Production of the Central Committee on Drilling and Production Practices, Division of Petroleum, American Petroleum Institute, (Dallas, Texas, 1942).

with petroleum deposits. Secondly, they conclude that regulation of the production rate is the most important factor governing efficient oil recovery, and thus regulation should be such that fields produce at their maximum efficient rate. Production in excess of reasonable market demand is defined by the Committee to be waste. Waste also results from improper well spacing patterns, improper well density, improper well completion, excessive encroachment of water or gas, and/or excessive production of water or gas. They conclude that state regulation has been somewhat effective in eliminating waste, but the regulatory system is in an evolutionary process, and as such, as knowledge of the laws of economics of engineering, and the physical laws of nature are uncovered, as well as the principles of land law, are known, the system of regulation should change and for the better to allow for a more efficient recovery system.

It is interesting to note presently that the individual systems of regulation have changed little over the lifetime of the regaltory agencies. Thus, the stagnation has resulted from political considerations and possible lack of economic concern for conservation. It certainly hasn't stagnated as a result of engineering or physical research, although little headway seems to have been made on discovering tertiary methods of recovery. The same can be said for the legal aspects, although again here the laws regarding regulation have traditionally been the football of the political game.

The principles the Progress Report sets out are rules governing the allocation within a pool and they are as follows: (1) that physical waste be prevented; (2) within reason, that each operator have an equal opportunity to recover an equivalent of the amount of recoverable oil underlying his property, however, to achieve this with undue waste; (3) that the method of allocation be simple enough to be able to be administered without imposing undue hardship on any particular operator.[16]

The Progress Report stresses the economics of conservation for the first time, particularly in regard to production rates and the allocation of production to meet the market demand. Further emphasis is laid at the necessity of controlling development costs, and the need to form an adequate capital budgeting system so that unnecessary expenditures in investment be prevented. The Progress Report states in regard to this point that: "In

---

[16]Ibid., pp. 7 and 8.

the interest of conservation, the drilling of unnecessary
wells should be discouraged or even prohibited. Further-
more, the drilling of unnecessary wells results in economic
loss which might otherwise be avoided."[17]

While the API undertook this Progress Report on
well spacing, the industry was not idle. In 1943, one year
after the issuance of the Progress Report, the industry came
back with their reply in the form of "The Joint Progress
Report on Reservoir Efficiency and Well Spacing."[18]
However, the Standard people took an entirely different
approach from the API study. The Joint Progress Report
is concerned basically with the various recovery mechanisms,
and the achievement of efficiency in recovery with the use
of each one of the combination of these mechanisms. Since
the mechanisms of recovery have been discussed in Chapter
the reader is referred to that section for a treatment of
the mentioned report.

Part III of the Joint Report does get into the

---

[17]Ibid., p. 11.

[18]Joint Progress Report on Reservoir Efficiency
and Well Spacing, The Committee and Reservoir Development
and Operation of the Standard Oil Company (New Jersey)
Affiliated Companies and of the Humble Oil and Refining
Company (New York, 1943).

problem of the fundamentals of well spacing, but the point
of view taken is that of flow of fluid through a reservoir
and it does little but reiterate the physical principles of
Muskat's 1939 work. They do conclude that time is the major
factor in determining the distance between wells as well as
other physical characteristics, but they do not take the
next obvious step and link time with economics of well
spacing. The Joint Report further emphasizes that well
location may well be more important than the actual number
of wells on the reservoir, but this leaves out all economic
considerations. They continue to argue that proper well
location is necessary for even and regular water and gas
movements of the various drive systems, and well location is
important when fault blocks or large lenses break up the
reservoir into separate units. The remainder of the Joint
Report is concerned with specific field studies, and the
results which they reach can be best summarized by the
statement in the Joint Report: "that in no field considered
by the committees was there apparent any physical relationship between well spacing and oil recovery from a reservoir
as a whole."[19] This, of course, is just the Miller-Higgins
conclusion in different words.

---

[19]Ibid., p. 64.

In the immediate post-war world the study on
spacing continued, particularly in 1945-46 a preliminary
study was conducted by Craze and Buckley. [20] In the study,
production information and statistical data for 103 oil
reservoirs were made available to the authors. The fields
were widely distributed geographically and a wide range of
reservoir and fluid characteristics were represented. The
well spacing on the reservoirs ranged from 2.6 to 65.6 acres
per well.

The Craze-Buckley approach to the spacing problem
was strictly one of analysis of the statistical data. In
order to determine if difference in well density or spacing
have a direct and recognizable effect on the amount of
ultimate recovery, they adapted a method of analysis by a
series of graphs. The graphs were designed to eliminate in
several steps the effects on recovery of such varying
factors as reservoir porosity, viscosity, permeability and
pressure decline. The method was strictly empirical. It
was a serious attempt to make refine the results of a study
to show differences in oil recoveries due to differences in

[20]R. C. Craze and S. E. Buckley, "A Factual
Analysis of the Effect of Well Spacing on Oil Recovery," API
Drilling and Production Practice, 1946; also see R. C. Craze
and J. W. Glanville, Well Spacing, Production Research
Division, Humble Oil and Refining Co., (Houston, Texas, 1955)

well density.

The results, according to Craze and Buckley, are for a gas-drive reservoir: "There is no apparent relationship between ultimate recovery and well spacing." And for water-drive reservoirs, they conclude: "There is no evidence in these data of a direct relationship between ultimate oil recovery and well spacing for the water-drive and fields studied." In summary, they show statistically that Miller and Higgins are correct in their 1939 article and little else.

The first attempt in the engineering literature at devising an economic model or an economic theoretical treatment of the problem was undertaken by Moyer.[21] He contends that by studying reservoir behavior and well pressure drawdown, it is possible to determine sand thickness and permeability. Once these are known, it is then possible to determine the economic producible reserves in a reservoir. After determining the economic ultimate recovery, he determines a productivity index, and then by simple calculation the best spacing for the reservoir in

---

[21]Vaughn Moyer, "Some Theoretical Aspects of Well Drainage and Economic Ultimate Recovery," Trans. AIME, (1948), Vol. 174, pp. 88-101.

question. Finally, he defines the economic rate of operation for a reservoir as: "The balance between total production costs and profits on a time basis."[22] He then
proceeds to define a curve which he calls the economic
ultimate recovery curve given in Figure 1. Continuing, he
develops a relationship between the productivity index and
the rate of production. This curve shows that a well will
have a higher indicated potential after a long production
period at a low rate than it would if operated at a high
rate for a comparable length of time. Thus, he derives the
production relation in Figure 2.

A further study of the equation that he uses for
well spacing also indicates that some knowledge can be
gained about the effect of the size of well bore on recovery. The equation he uses is given by:

4
$$P_e - P_w = \text{constant} \times \frac{Q}{Kh} \cdot \frac{\mu_o B_o}{K_o}$$

where he has defined:

P_e = outer well pressure in psia;

P_w = inner well pressure in psia;

Q = total volume STB/day;

K = permeability;

---

[22]Ibid., p. 96

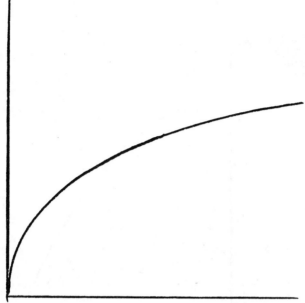

Figure 1   Economic Ultimate Recovery Curve

Note: This graph shows that as we increase the
             well spacing we can anticipate greater
             economic ultimate recovery per well.

             There is some upper limit on the well
             density factor which can be determined
             from the economic life of the reservoir,
             and thus the curve flattens out as it
             reaches this limit.

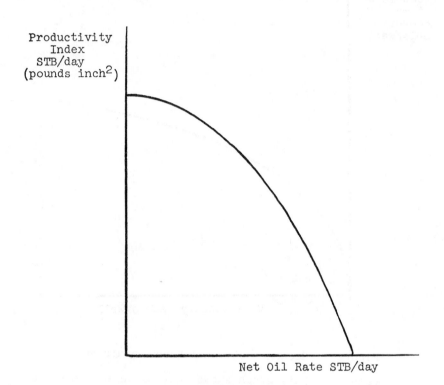

Figure 2   Production-Oil Relation

Note: For this graph see Vaugh Moyer, "Theo-
             retical Aspects of Well Drainage and
             Ultimate Recovery," Trans. AIME, Vol.174,
             (1948), pp. 96 and 97. It relates the
             productivity index later defined as J to
             the net oil production rate, and thus
             defines a production possibility set.
             By varying the productivity index, and
             thus the life of the reservoir, we obtain
             an array of production possibility sets.

K_O = permeability of the oil phase

N_O = visiosity of the oil phase

B_O = formation volume factor bbl/bbl

h = sand thickness

The conclusion derivable from this is that ultimate economic recovery may be increased slightly by drilling a large hole. Also, the productivity index is slightly increased indicating a shorter producing life. This total effect decreases as the permeability of the reservoir increases. The conclusion derived is that it is important to consider the well bore when determining the production rate and it has a little effect on well spacing but is not a major determinant.

Moyer's study is really the first study relating economic ultimate recovery to the rate of production to well spacing, and realizing the interdependence of the three factors and their economic content.

In a last attempt to prove Cutler's work fruitful, Tomlinson, in 1950, treats the subject of well spacing in great detail as a mathematical and physical flow of fluids problem. His purpose is to prove once and for all the

validity of Cutler's Rule.[23] The work only need be mentioned here since the Miller-Higgins study showed beyond a
shadow of a doubt the invalidity of Cutler's Rule.

In 1951, Bartram presented a report entitled "Well
Spacing From a Geologist's Viewpoint."[24] In this paper, he
refers to the Craze-Buckley study and to the 1950 Tomlinson
paper and attempts to reanalyze both papers and their data
from the geologist's point of view. In the final analysis,
Bartram agrees with the Craze-Buckley findings, that well
density is not an important factor in securing maximum
recovery from an oil reservoir. However, concerning the
spacing problem and its solution by statistical methods
Bartram comments: "Statistical analyses have not settled
the question, and the argument continues. In the following
paragraphs, some of the previous statistical analyses will
be discussed (Craze-Buckley) and some new data presented.
This will show that different investigators can draw
opposite conclusions from essentially the same data, and

[23]C. W. Tomlinson, "Well Spacing and Use of
Energy" Research and Coordinating Committee, Interstate Oil
Compact Commission, 1950.

[24]John G. Bartram, "Well Spacing From a Geologist's Viewpoint" Research and Coordinating Committee,
Interstate Oil Compact Commission, April, 1951.

that statistics can probably never settle the well spacing
problem."[25]

In treating the subject of well density, Kaveler
in his 1950 report cites neither principles, theories, or
statistics. Rather, he cites the conclusions which either
he or other students of well spacing have found.[26] Thus,
his analysis is qualitative and can be best summarized by
hs own words: "Petroleum technology teaches, and reliable
field data is rapidly accumulating as substantial proof that
within the range of well spacing usually encountered in the
oil fields of the United States, that is, in the range of
one well per five acres up to one well to 40 or 80 acres,
the ultimate recovery of petroleum from a pool is substantially independent of well spacing or number of wells
drilled."[27]

In another report on well spacing, "More Wells --
More Oil"[28] by Kaveler, he cites production performance
records from six pools in Oklahoma that were originally

[25]Ibid., p. 79.

[26]H. H. Kaveler, "Some Considerations in Regulation of Well Spacing," Research and Coordinating Committee,
Interstate Oil Compact Commission, (1950).

[27]Ibid., p. 63

[28]H. H. Kaveler, "More Wells -- More Oil," The
Petroleum Engineer, August-September, 1950.

partly developed on 40 acre spacing plans, and after establishment of definite production trends, further developed by infill wells on a 10 or 20 acre basis. He plotted production data for the separate pools on Log-Log graphs. The production curves showed very close agreement between the ultimate expected recovery from the pools while producing from 40 acres wells and that expected after additional wells were drilled. This was true even when the later drilling expanded to the pool boundaries. The graphs definitely show the interference effect on production from older wells by diversion of oil to later infill or field extension wells. The study further furnished substantial proof that these closely spaced pools were not effective in bringing about greater recovery than wider spacing would have done. It also furnishes evidence that where faulting or other obstructions do not interfere, oil will travel horizontally over substantially greater distances to reach the well outlets than previously had been believed.

Kaveler's statement of principles established by petroleum engineering technology in respect to well spacing is worthy of quote. They are:

"1. Define the extent of the pool and the limits of production.

2. Insure that every segment of the pool

productive of petroleum is penetrated by
at least one well.

3. Insure sufficient wells, after consideration
of geological and reservoir characteristics
of each pool, to permit the maximum bene-
ficial use of either natural or artificial
sources of reserve energy.

4. Insure sufficient number of wells to maintain productive capacity at rates equal to
or in excess of market demand for petroleum."[29]

In the matter of studies of physical principles

involved in flow of fluids through porous media, and the

application of the theories to well density problems, Muskat

has been the most thorough of any work in this area.[30] His

work considers in more detail the economic aspects of well

spacing, and it is relevant to review here his important

work. His analysis of the well spacing problem is a growth

of Moyer's analysis. His consideration of the physical

principles are first included in his 1949 work, and he

presents in great detail the work which he had completed

on the flow of fluids in a reservoir.

However, in 1949, Muskat realized and discussed

in more detail the economic considerations that needed to be

[29]Ibid., p. 61

[30]M. Muskat, Physical Principles of Oil Production, Chapter 14, (New York, The MacMillan Co., 1949),
pp. 810-904.

considered in analyzing the well spacing problem. First,
there is the recognition of the economic problem: "The
number of wells used ... should be determined mainly by the
economic balance between the cost of drilling and operation
and the value of the additional recovery that may be
obtained from additional wells through the achievement of
greater geometrical sweep efficiency."[31] Secondly, he
recognizes the importance of the time factor in the recovery
process of a reservoir and indicates: "the time factor ...
should be considered in relation to the danger of accelerated pressure decline resulting from excessive withdrawal
rates."[32] The well spacing problem then is essentially
made up of two component questions: first, how does the
physical ultimate oil recovery vary with the well spacing?;
and secondly, how does the production rate per well versus
time vary with the well spacing? The answers to these
questions will give the optimal well spacing configuration.

The first question pertains solely to the purely
physical relationship between oil expulsion and well
spacing, the distance or radius of drainage. These physical considerations have been considered in Chapter 3 where

[31]Ibid., p. 820
[32]Ibid., p. 820

we consider the purely physical relationship of oil and gas recovery and the flow of fluid through porous media.

Here we are more interested in pursuing to a greater degree the second question which has, and introduces, the economic aspects of the problem. The first fact to consider is that even if the physical ultimate recovery from two different reservoirs be identical in every respect, the economic factors affecting ultimate recovery may be different, namely the production rate versus time may be different and thus, we have a different problem and different solution for identical reservoirs.

For gas-drive units, Muskat in a series of papers concludes that from laboratory experiments and from collected data: "The physical ultimate recovery from uniform solution gas-drive fields may be considered as independent of the well spacing."[33]

As for the economic ultimate recoveries, these can be determined from the production decline curve. The

[33]M. Muskat, "Principles of Oil Well Spacing," Trans AIME, (1940), Vol. 135, p. 37.

_____ and M. W. Meres, "The Flow of Hetrogeneous Fluids Through Porous Media," Physics, (1936), Vol. 7, p. 325.

integral under such a curve to the time when the production

rate has fallen to the abandonment limit will give the

ultimate recovery. If such curves were constructed and

integrals evaluated for various well spacings, the variation

of the economic ultimate recovery with the well spacing

would be obtainable directly. Since the production rate per

well will be given in Chapter 3, under standard operating

assumptions, as:

5
$$q_{sc} = \frac{7.08kh\,(Pe - Pw)}{Bo\mu\left[\ln\dfrac{r_e}{r_w} - \dfrac{1}{2}\right]}$$

we are then permitted to plot this production rate against

cumulative production, and also against cumulative recovery

as a percent of the pore space in the reservoir as in

Figures 3 and 4.

In actually carrying out the economic aspects of

the well spacing problem, the production decline curve

plays an important role. The interest factor associated

with the operating life may actually control the ultimate

well spacing which will yield maximum profit. If the

interest rate at any point in time is "i", and if the pro-

duction per well during the time interval is $q_{sc}t$, then by

simple discounting, it is possible to find the future worth

of oil production per well by the formula:

Production
Rate in STB

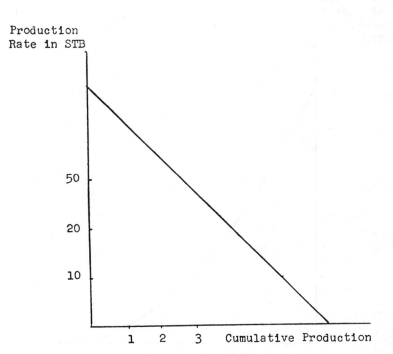

50

20

10

1   2   3   Cumulative Production

Figure 3   Economic Ultimate Recovery vs.
Cumulative Production

Production
Rate per
Reservoir
STB/day

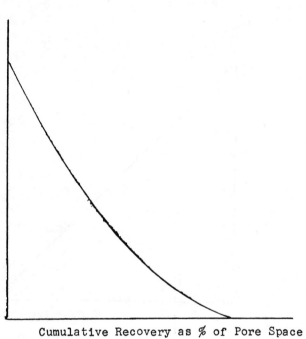

Cumulative Recovery as % of Pore Space

Figure 4   Economic Ultimate Recovery vs.
Cumulative Recovery as a % of
Pore Space

6
$$FWOP = \sum_{t=1}^{T} q_{sct} \frac{(V - C)}{(1 + i)} t$$

where V is net revenue per barrel of oil and C is the
operating cost per barrel of oil and q_{sct} is expressed as
the total production in STB for the t periods. If the
drilling and operating costs are c, revenues obtained are r,
and investment costs be I per well, then the present value
of the future profits from a series of n wells is given
by:

7
$$PV = n \left[\sum_{t=1}^{T} q_{sct} \frac{(r - c)}{(1 + i)} t - I \right]$$

8
$$PV = \frac{A}{Ao} \left[\sum_{t=1}^{T} q_{sct} \frac{(r - c)}{(1 + i)} t - I \right]$$

where A is the productive area and Ao is the drainage area
per well assuming all wells are drilled simultaneously.

From this present value formula it is possible to
obtain the operating income curve as a function of Ao, the
well spacing, as is given in Figure 5 . Operating income
will be a maximum at zero well spacing, since the reservoir contains the maximum ultimate recovery before the
wells are actually drilled. As the well density decreases,

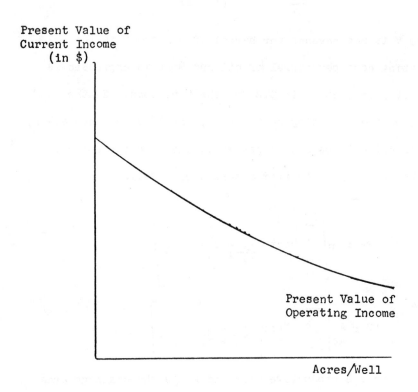

Figure 5 Operating Income with Variations
in Well Spacing

the operating income curve falls due partly to the decreasing economic recovery with increasing well spacing, but also due to the fact of increasing operating life, to abandonment, and to increasing discount as the well spacing is increased.

The well investment curve in Figure 6 is the difference between the operating income and the well investment. It has a maximum value where the slope of the operating income curve is equal to that of the well investment curve. Thus, optimum economic well spacing is seen in Figure 8 where these two slopes are equal.

Obviously the assumptions made about the physical characteristics of the reservoir will affect both the operating income curve, the well investment curve, and thus, the net profit curve and thus, the optimum economic well spacing. As an example, suppose that the permeability of the reservoir studied declines. This will affect the production rate, causing it to decline thus causing the operating income curve of the reservoir to become more flat. In absolute value of the economic ultimate recovery, the change in the income curve will be such that the new point of optimum economic well spacing will be such that there will be more wells drilled. Thus, we can establish

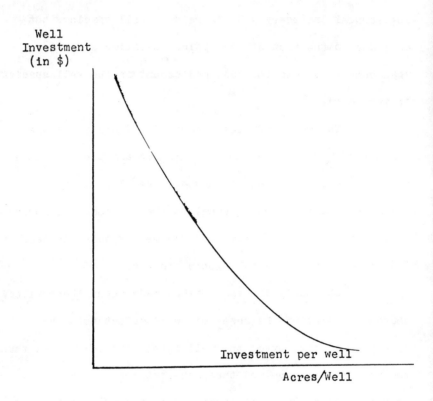

Well
Investment
(in $)

Investment per well

Acres/Well

Figure 6 Well Investment with Variations
in Well Spacing

53

Net Profit
(in $)

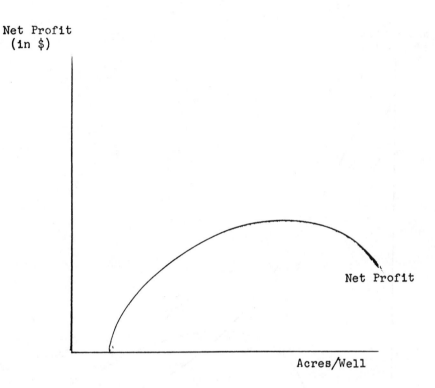

Net Profit

Acres/Well

Figure 7   Net Profit with Variations for
Well Spacing

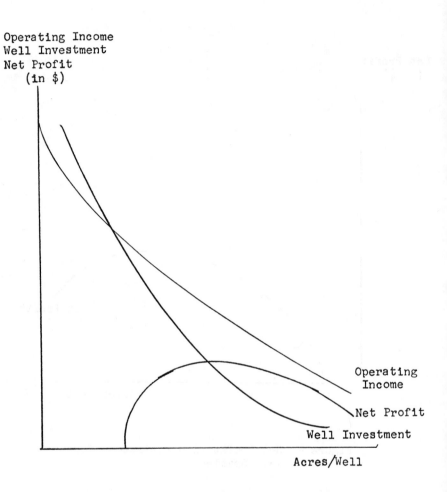

Operating Income
Well Investment
Net Profit
(in $)

Operating
Income

Net Profit

Well Investment

Acres/Well

Figure 8   Operating Income, Well Investment, and
Net Profit and Determination of Optimum
Economic Well Spacing

relationships between behavior of K and h and well spacing.
As K and h increase, the well spacing will increase like-
wise so that $\frac{\partial Ws}{\partial Ks} > 0$ and $\frac{\partial Ws}{\partial h_s} > 0$. A shift toward lower
well densities and higher optimum well spacing will result
from increased well investment cost. Higher abandonment
rates, that is, the time when the well is abandoned becomes
shorter, will also mean wider well spacing patterns to
achieve maximum profit.

To derive a specific relationship between profit
and well spacing, Muskat assumes a production rate that will
decline exponentially with time. Also, assuming that the
ultimate physical recovery is independent of the well
spacing pattern, then the rate of production can be represented as:

9 $$q_{sc} = q_{sc1} \cdot \exp^{-\alpha kt/re^2}$$

where q_{sc} is the initial production rate under standard
operating techniques, α is a constant determining the
ultimate physical recovery, and r_e is the drainage radium
per well corresponding to the well spacing pattern.[34] The
physical ultimate recovery per well is given by:

---

[34]Ibid., p. 828.

10 $$\text{Pur} = \frac{q_{sci}\, r_e^2}{k\,\alpha}$$

and the cumulative recovery to the time when the production
rate has become q_{sca} is given by:

11 $$\text{Pcul} = \text{Pur}\ 1 - \frac{q_{sca}}{q_{sci}}$$

If q_{sca} is considered as the abandonment rate,
then the economic ultimate recovery curve will increase
linearly with decreasing abandonment rate as can be seen
from Figure 9.

Setting this productive relationship into the
framework of the calculation of present worth of profit, if
the production rate in time period (t-1) and time period t
is represented by equation

12 $$q_{sct} = \text{Pur}\ .\ \exp - (t-1)\alpha kt/r_e^2 \left[\exp -\alpha kt/r_e^2\right]$$

then the present worth of product discounting over time is:

13 $$\sum_{t=1}^{T} \frac{q_{sct}}{(1+i)}\ t = \frac{\text{Pur}\ (1 - e^{-g})}{1 + 1 - e^{-g}}\left[\frac{1 - e^{-g}}{(1+i)}\right]$$

where g has been defined as $\alpha kT/r_e^2$.

Again assuming that the revenues from oil are
given by r per STB and that costs are c per STB, and that
they remain constant throughout the life of the reservoir,

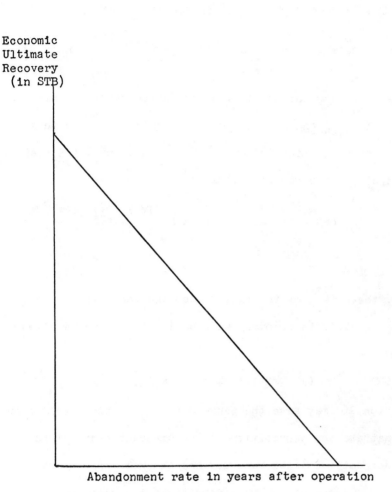

Abandonment rate in years after operation

Figure 9   Economic Ultimate Recovery vs.
Abandonment Rate

then the present worth of net income can be given by

14 $\qquad PV = A \dfrac{Po\ (r - c)\ (1-e^{-g})}{1 + i - e^{-g}} \left[1 - \dfrac{q_{sca}}{q_{sci}\ (1 + i)^{t}} \right]$

where Po is now the physical ultimate recovery per unit area. If q_{sca} and q_{sci} become very small, then the term in brackets in equation 14 approaches 1, and thus, the present worth of profits can be written as:

15 $\qquad$ Present Worth of Profits $= A \left[\dfrac{Po(r-c)(1-e^{-g})}{1+i - e^{-g}} \right]$

$\qquad\qquad\qquad\qquad\qquad\qquad\qquad - \left[\dfrac{I}{Ao} \right]$

Thus, the effect of the time factor and variation of the economic ultimate recovery with well spacing can be given by the term:

16 $\qquad (1 - e^{-g})/(1 + i - e^{-g})$

Equation 16 may have the range 0 to 1/1+i which would then demonstrate the sensitivity of the discount factor i to the economic ultimate recovery and the well spacing pattern.

$\qquad$ Also, as the well density is increased (decreased), g will increase (decrease) and so will net operating income, and thus the curve representing net operating income will flatten (become more steep).

17 $\qquad \exp^{-\ q_{sci}t/PoAo} = 1 + i + B - \sqrt{B^2 + 2B\ (1+I)}$

where B $= \dfrac{q_{sci}\ i(r-c)^t}{2I}$. Thus, in summary q_{sci} will depend

on the permeability of the reservoir. Likewise, profit will

decrease as the well cost investment increases. However,

spacing will increase as the net revenue per STB increases

or as the average recovery Po increases.

Every petroleum textbook since the Muskat study

includes a short section dealing with the economics of the

well spacing problem. However, very little literature on

the economics of well spacing developed in the engineering

journals, save Miller and Dynes study on proper well spacing

abroad.[35] Their analysis can be briefly reviewed by saying

that they apply the Muskat techniques to the Mid-East fields

and determine that wide spacing maximizes the present value

of profits of fields in that area. A wider spacing is

permitted, of course, partially because of the geological

factors which Miller-Dynes leave out of their study, i.e.,

the absence of faults and lenses which predominate the

structure of reservoirs in the United States.

---

[35]C. C. Miller and A. B. Dynes, "Maximum
Reservoir Worth-Proper Well Spacing," Trans. AIME, (1959),
Vol. 216, pp. 334-340.

Field Studies

While the above analysis has attempted to present
only the theoretical analysis of well spacing in its historical environment, to the present day, this is not to
infer that the theoretical studies are the most important
aspect or the only aspect of the problem of well spacing
studied. Much work has been done, and some of the more
important field studies have been referred to in the above
analysis, to determine the optimal spacing from a statistical analysis.

It is not our concern here to review all this
literature, for the techniques used vary widely and much of
analysis depends directly on the particular field being
studied. Our concern has been with general principles, and
thus, we feel if individual analysis of a specific field is
desired, or the historical analysis of how fields have been
studied in the past is of interest to the reader, the
reader is referred to that analysis.

Summary

The general treatment of well spacing in the
engineering literature has been reviewed. Studies not
included in this review have been the more recent trends in
the engineering literature to treat the physical problems

in greater depth, that is unsteady-radical flow, logging, mudding, heat control of the liquid, and the entire set of problems surrounding secondary and tretiary recovery. For these more technical studies, the reader is referred to the technical journals in petroleum engineering.

It is important to note that since the Miller-Dynes study in 1959, the engineering profession has placed the problem of well spacing on the doorstep of the economist to solve. How has the economic profession reacted to this shelfing of responsibility? Reviewing the literature which has evolved in economics concerning the spacing problem is not a particularly laborious task. Economists have ignored the problem of well spacing. However, economists have been concerned with an allied problem, that of conservation, and thus bits and pieces of analysis have filtered into the literature. Before considering these economic developments in greater detail, which we do in Chapter 5, it is necessary for the analysis to summarize the physical and engineering considerations so that the reader will have some idea as to what reserves mean and how they are calculated, and likewise what is the current state of the art like to discuss the physical problems and relationships necessary to complete understanding of the well spacing problem.

CHAPTER 3

CALCULATION OF GAS AND OIL RESERVES

Introduction

While well spacing is not solely a function of
the physical conditions of the reservoir as was previously
believed, these physical and engineering considerations do
enter when determining an optimal investment policy.
Basically there are two aspects of the physical and engineering problem that need to be considered; that of calculation of reserves and secondly, the problem of flow of fluid
through a porous media.

Concerning the calculation of the potential
reserves, there are many possible methods of calculating
this stock; the volumetric method, material balance method,
and production decline method being the most common. Here
we shall describe the first two in some detail. We shall
find that in order to calculate reserves, we need to
determine the volume of the reservoir, the porosity, drive,
and pressure behavior.

For the flow of fluids through a porous media, we shall need to consider viscosity, compressibility, drainage radius, pressure build-up, and flow systems of the reservoir.

This treatment is not to be considered as an extension of the engineering literature, but rather a presentation of the physical principles that need consideration when considering the well-spacing aspects of the investment and production policy for a reservoir. We intend only to present the material so that the reserves and flow aspects of the well spacing problem can be considered in their proper physical aspects.

Reserves

Our first problem is that of determining the reserves of the well-find, given that the exploration activity have indicated a producing well in order that the rate of production may be defined. Reserves are defined as the amount of hydrocarbons in a field, gas, oil, or gas-oil. The measurement of reserves usually indicate that gas or oil may be recovered at an economic rate using only the natural reservoir forces, that is, these reserves recovered by natural force are given the name "primary reserves." Reserves resulting from secondary recovery, and thus called

"secondary reserves," will not be considered in this
analysis since we are concerned only with well-spacing,
investment, and production necessary to economically recover
the primary reserves.

In calculating reserves, it is important to note
that the major concern is a "relative" measure of reserves,
that is, arriving at a figure that is consistent with the
data given at any particular moment in time. If the data
changes, i.e., more data becomes available, either engineering or geological, a more accurate estimate of primary
reserves should be made in light of the data recovered.

There are many methods of estimating gas and oil
reserves. Here we will be concerned with the more basic
methods are:

(1) Volumetric estimation;

(2) Material balances equation;

(3) Production decline curves;

(4) Comparison of the reserves with those of
      similar geological and reservoir properties;

(5) Comparison of field data from the same
      formation in other fields.

Reservoirs are usually classified in accordance
with the fluid which they produce. The more common types

of categories are: (1) oil; (2) gas and gas-condensate; and (3) volatile oil. In this study, we shall be concerned with all but volatile oil reserves.

The volumetric estimation of either oil or gas consists primarily of determining the volume of rock holding hydrocarbons, the total voids in such rock, the volume percentage of the voids containing hydrocarbons, and what percentage of these hydrocarbons are economically recoverable as primary reserves. Reserves are usually expressed in either stock tank barrels (STB) for liquid oil, or million standard cubin feet (MMSCF) for gas.

The standard formula used for calculating gas reserves is slightly more complicated than the one used for calculating oil reserves, primarily because the gas reserves are somewhat affected by the source of the gas. Throughout the depletion of a gas or oil reservoir, there will most likely be several different sources of the oil or gas which have to be considered.

The various types of gas one might find in a reservoir are:

(1) Non-associated gas-free gas not in contact with liquid hydrocarbons.

(2) Associated gas-free gas in contact with

liquid hydrocarbons.

(3) Dissolved gas -- gas in solution with

liquid hydrocarbons.[1]

In the case for oil, it is simply crude that interests us.

Usually to simplify matters, when gas reserves
are being estimated no effort is made to distinguish between
associated gas and dissolved gas; only total gas reserves
are estimated. Thus, we shall ignore the difference between
these two sources of gas and give it the name associated gas
meaning both associated and dissolved gas.

The volumetric equation for the calculation of
non-associated or associated gas reserves is given by:

18
$$V_i - V_f = 43,560 \ Ah \ \emptyset \ (1-S_w) \ R.F. \ \frac{T_B}{P_B T_r} \left[\frac{P_i}{Z_i} - \frac{P_f}{Z_f} \right]$$

where:

V_i = volume of gas initially in place in (SCF);

V_f = estimated volume of gas at abandonment in (SCF);

43,560 = number of square feet per acre or number of cubic feet per acre foot;

A = area;

---

[1]Charles A. Hawkins, Cost Problem in the Field
Price Regulation of Natural Gas, unpublished thesis, (Purdue
University, 1964), pp. 2-3.

h $=$ net pay thickness;

$\emptyset$ $=$ porosity; fraction of voids in total rock matrix;

S_W $=$ volume fraction of the voids filled with interstitial water;

R.F. $=$ fraction of the gas in place that probably will be recovered;

T_B $=$ base temperature 520° R;

P_B $=$ base pressure in psia;

T_r $=$ reservoir temperature in ($^\circ$R);

P_i $=$ initial pressure;

P_f $=$ estimated pressure at abandonment in psia;

Z_i $=$ compressibility factor at P_i and T_r;

Z_f $=$ compressibility factor at P_f and T_r;

$^\circ$R $=$ degrees rankine $(^\circ R)^2$;

This equation 18 above accounts also for the gas left in the ground at the time of abandonment. In this case the recovery factor accounts for the inefficiencies incurred in the depletion of the reserve.

The recovery factor (R.F.) is somewhat arbitrary and does little more than estimate the probable error to be incurred in calculating reserves. It is common to use

[2]Many of these formulas are taken directly or indirectly from Donald Katz, et. al. Handbook of Gas Engineering (New York, McGraw Hill, 1959), Chapter 10-11, pp. 403-464.

values from .75 to .80 for slimy and shaly sands, with
proportionately higher values for cleaner more permeable
sands.

Volume

One of the necessary steps in solving the volumetric equation is the estimation of the number of acre-
feet of the reservoir containing gas-condensate and oil.
The only reliable method in use is to draw an isopachous[3]
map of each phase of the well as the test data become
available.

The first step is to determine the probable pay
thickness of each phase in the well. The tools necessary
for the determination of this aspect are the various types
of log analysis available in well drilling, i.e., electric,
radioactive, nuclear, etc., along with either core analysis
and geological sample descriptions. The problem of thickness is compounded by many difficulties. One of the more
important is the location of contacts or other productive
regions in the reservoir. The change of phase with levels
is not instantaneous and there is often considerable

---

[3]Isoporchous is defined as: An isogram that
connects points of equal thickness of a particular geological stratum formation or group of formations.

latitude, plus the fact that a given contact may vary in each well drilled. The usual procedure is to use the cores available during the test drill(s) and to correlate them with the well logs and geological sample available.

The relative thickness of gas or oil sands in contact may be estimated from gas-oil data just prior to completion. There are three equations which are commonly used to determine the volume of a productive zone. The first is that of the frustum[4] of a pyramid which is given by the equation:

19
$$V_b = \frac{h}{3} A_n + A_{n+1} + \sqrt{A_n + A_{n+1}}$$

where we define:

V_b = bulk volume in acre feet;

h = interval between isopach lines in feet;

A_n = area enclosed by lower isopach lines in acres;

A_{n+1} = area enclosed by upper isopach lines in acres.

This equation 19 is used to determine the volume of gas or oil between the successive isopach lines and the volume is just the sum of these two separate volumes, i.e., the sum

---

[4]The solid figure formed when the top of a cone or pyramid is cut off by a plane parallel to the base.

of the upper and lower isopach lines. The second method is that of the trapezoidal rule whose equation is given by:

20 $$\triangle V_b = \frac{h}{2} (A_n + A_{n+1})$$

or if the reservoir is a series of successive trapezoids it then 20 becomes:

21 $$\triangle V_b = \frac{h}{2} (A_o + 2A_1 + 2A_2 + \ldots + 2A_{n-1} + A_n)$$
$$+ T_{avg}A_n$$

where V_b, and h are defined as before and:

A_o = area enclosed by the zero isopach line in acres;

A_1 = area enclosed by the first isopach line in acres;

. . . . .

. . . . .

. . . . .

A_n = area enclosed by the nth isopach line in acres;

T_{avg} = average thickness of the isopach lines in feet.[5]

Again we determine the volume of gas or oil in the zero trapezoid, first trapezoid, etc., and the volume is merely the sum of the separate trapezoids.

The third rule that can be used to determine the

---

[5]Donald Katz, et. al., Handbook of Gas Engineering, (New York, McGraw Hill, 1959) p. 411.

bulk volume is given by Simpson's rule which states that:

22 $$\Delta V_b = \frac{h}{3}(A_0 + 4A_1 + 2A_2 + 4A_3 \ldots$$
$$+ 2A_{n-2} + 4A_{n-1} + A_n) + T_{avg}A_n$$

where everything is defined as it was in equation 21 . The difference between Simpson's rule and the trapezoidal rule is that Simpson's rule is more accurate for irregular reservoirs. However, one of its drawbacks is that is is far more tedious to use and to work out. However, for calculating the volume, any one of the above methods are adequate and the particular method used should be determined from the isopach map arrived at from the well logs and geological analysis.

In many gas and oil reservoirs, particularly during the period we are concerned with -- the development period -- the bulk volume is not known. In this case then it is better to place the reservoir calculations on a unit basis, normally on a one acre-foot of bulk reservoir rock. Then one unit of bulk reservoir rock contains:[6]

$$C_w = 43,560 \; \emptyset \; (S_w) \; \text{Connate water}$$

$$R_{gv} = 43,560 \; \emptyset \; (1-S_w) \; \text{Reservoir gas volume}$$

---

[6]Connate can be defined as: water entrapped in sediments at the time of their deposition: originating at the same time as adjacent interminized materials.

$R_{pv} = 43,560 \, \emptyset$ Reservoir pore volume.

The initial gas in place given in SCF in the unit is given by:

23 $\qquad G_i = 43,560 \, \emptyset \, (1-S_w) \, V_{gi} \quad$ SCF/ac.ft.

where we define:

$\quad G_i =$ initial gas volume in SCF;

$\quad \emptyset =$ porosity factor expressed as a fraction of the bulk volume;

$\quad S \; =$ connate water as a fraction of pore volume;

$\quad V_{gi} =$ gas volume factor in SCF per cubic foot.

The gas in the reservoir at abandonment time then is given by:

24 $\qquad G_f = 43,560 \, \emptyset \, (1-S_w) \, V_{gf} \quad$ SCF/ac.ft.

The unit recovery is the difference between the gas initially in place G_i and the gas left at abandonment G_f. Combining, we get that the unit recovery is given by:

25 $\qquad U_r = 43,560 \, \emptyset \, (1-S_w)(V_{gi}-V_{gf}) \quad$ SCF/ac.ft.

where $U_r =$ unit recovery. The fractional recovery or recovery factor expressed as a percent of the gas initially is given by:

26 $\qquad R. \, F. \; = \; \dfrac{100(G_i - G_f)}{G}$

$\qquad R. \, F. \; = \; \dfrac{100(V_{gi}-V_{gf})}{V_{gi}}$

These recovery factors are valid for any given reservoir
provided there are no drains in the reservoir to adjacent
units.

Water Drive

Under initial conditions, one unit of bulk rock
contains:

$$C_w = 43,560 \; \emptyset \; (S_{wi}) ft^3 \qquad \text{Connate water}$$

$$R_{gv} = 43,560 \; \emptyset \; (1-S_{wi}) ft^3 \qquad \text{Reservoir gas volume}$$

$$S_{ug} = 43,560 \; \emptyset \; (1-S_{wi}) V_{gi} \qquad \text{SCF} \quad \text{Surface units gas.}$$

In many reservoirs under water drive, the pressure suffers
an initial decline after which water enters the field at a
rate equal to or close to the rate of production, thus
causing pressure to stabilize. In this case, the stabilized
pressure is the abandonment pressure. If V_{gf} is the gas
volume factor at the abandonment pressure and S_{gr} is the
residual gas saturation expressed as a fraction of pore
volume after water enters the reservoir, then:

$$W_v = 43,560 \; \emptyset \; (1-S_{gr}) cu/ft \qquad \text{Water volume}$$

$$R_{gv} = 43,560 \; \emptyset \; (S_{gr}) cu/ft \qquad \text{Reservoir gas volume}$$

$$S_{ug} = 43,560 \; \emptyset \; (S_{gr})(V_{gi}) SCF \qquad \text{Surface unit gas,}$$

and the unit recovery is the difference between the initial
and the surface units of gas:

27    $U_r = 43,560 \; \phi \; (1-S_{wi})V_{gi}-S_{gr} \cdot V_{gf}$

The recovery factor then becomes:

28    $R.F. = \dfrac{100 \; (1-S_{wi})V_{gi}-S_{gr} \cdot V_{gf}}{(1-S_{wi})V_{gi}}$

In general, it is safe to say that in gas recoveries by water drive, the rate of recovery is lower than by volumetric depletion, i.e., by gas drive. However, it does have an advantage in that a higher flow rate can be maintained as well as a higher well head pressure compared with gas drive, this due of course to the maintenance of pressure by the influx of water.

Porosity

In describing the volumetric equation 18, we indicated that porosity represented that portion of the total rock volume that contains fluids or gas, or sands that contain the same. This introduces a problem into the analysis since it is difficult, if not impossible, to distinguish between that porosity which exists along with permeability and thus permits production, and that porosity which does not have permeability and thus does not emit production. Thus, the porosity which emits production we shall term effective porosity and we shall compare it to total porosity.

Further complications exist in studying porosity because it occurs in crevices, vugs,[7] and fractures, as well as in intergranular and intercrystalline spaces. By virtue of their characteristics, cores, electric logs, radioactive logs, etc.,[8] show primarily intergranular and intercrystalline porosity. In sandstone, this presents no problem. However, in reservoirs other than sandstone, and this represents a large portion of reservoirs, this does and can present innumerable problems.[9]

The porosity of a well normally varies somewhat from producing section to producing section, and the most common means of circumventing this problem is to determine some average well porosity. Core analysis shows porosity for foot-intervals, while the logging interval has to be broken up into several sections and each section analyzed separately. To determine average well porosity, the simplest method would be:

29
$$a_\phi = \frac{h_1\phi_1 + h_2\phi_2 + h_3\phi_3 + \ldots + h_n\phi_n}{h}$$

---

[7]A vug is a small unfilled cavity in a lode or in a rock.

[8]Sylvain J. Pirson, Handbook of Well Log Analysis, Englewood Cliffs, N.J., Prentice-Hall, 1963) various parts.

[9]Ibid., p. 63.

where we define:

a_ϕ = average well porosity;

$h_1...h_n$ = number of feet of formation having a corresponding porosity of $_1, ..., _n$;

$h = h_1 + h_2 + h_3 + ... + h_n$.

Another problem which exists, and thus enters the analysis, is that water is sometimes found in the gas or oil bearing parts of the reservoir above the transition zone. This water is commonly called connate or interstitial water. It is important because the water reduced the amount of gas or oil available for recovery mainly by affecting and reducing the amount of pore space available for the fluid to be in. Generally, the water is not uniformly distributed throughout the reservoir, but varies with the permeability and the lithology, and with the height above the free water. There are a number of methods available for measuring this connate water. Schilthuis, for example, describes the only direct method of measuring this water. His methodology con- sists of coring the formation with an oil-based solution, and in this manner measuring the relation between absolute permeability and connate water.[10] There have also been

----

[10]Ralph J. Schilthuis, "Connate Water in Oil and Gas Sands," Trans. AIME., (1938), Vol. 127, pp. 199-214.

developed many indirect methods of measuring the extent of
connate water, but this takes us too far a-field and into
the engineering problem of connate water in gas and oil
reservoirs.[11]

Pressure

In what we have seen so far, and what we shall see
later, reservoir pressure indeed does play a most important
role in well analysis. In view of the fact that pressure
varies throughout a field, it is necessary to arrive at
some sort of an average pressure for a field. One way to
treat pressure is to plot isobaric[12] maps by taking the
pressure readings at development and experimental wells at
periodic intervals. These lines superimposed on an iso-
pachic map can give us some idea of the pressure range and
distribution in a field. The various measures of pressure
in a field are given by:

30 $$\text{Well average} = \sum_{o}^{n} \frac{P_i}{n}$$

---

[11]Ibid., as well as: Leverett; Bruce and Welge;
MacCallough; Albaugh and Jones; Messer; Archie; in various
issues of Trans. AIME., petroleum engineering section.

[12]Isobaric is defined to be: A line on a map con-
necting points in a global structure which have equal
pressure.

31 $$\text{Areal average} = \sum_{o}^{n} \frac{P_1 A_1}{A_1}$$

32 $$\text{Volumetric average} = \sum_{o}^{n} \frac{n P_1 A_1 h_1}{\sum_{o} A_1 h_1}$$

Materials Balance[13]

Up to the present we have been concerned with the factors involved in measuring initial gas or oil in place and have given the volumetric method of calculating reserves as well as a method based on a unit, one acre foot of bulk productive rock given from a knowledge of the porosity factor and connate water. Thus, to calculate the initial gas or oil in place on any particular portion of a reservoir it is necessary to know the bulk volume of that particular portion of the reservoir. In most actual cases, however, the porosity, the connate water and bulk volumes are not known with any precision, and thus, the methodology described has very drastic limitations. In these cases then it is useful to use a process known as the materials balance approach to calculate the initial gas or oil in the reservoir. It is important to note that this method is

---

[13]See D. Katz et. al. for a discussion of this section; Chapter 12, pp. 482-501 are the most relevant.

applicable to the reservoir as a whole because of the migration of the fluid from one portion of the reservoir to another in both the volumetric and water-drive reservoirs.

The conservation of mass may be applied to the case of gas or oil reserves to give the most general materials balance equation:

33
$$\begin{bmatrix} \text{weight of} \\ \text{fluid produced} \end{bmatrix} = \begin{bmatrix} \text{weight originally} \\ \text{in reservoir} \end{bmatrix}$$
$$- \begin{bmatrix} \text{weight remaining} \\ \text{in reservoir} \end{bmatrix}$$

where the composition of the production is constant. The standard cubic feet both produced and remaining in the reservoir are directly proportional to the mass, and we can write the materials balance equation also in terms of standard cubic feet:

34
$$\begin{bmatrix} \text{SCF produced} \\ \text{from reservoir} \end{bmatrix} = \begin{bmatrix} \text{SCF initially} \\ \text{in reservoir} \end{bmatrix} - \begin{bmatrix} \text{SCF remaining} \\ \text{in reservoir} \end{bmatrix}$$

If V_i is the initial gas pore volume in cubic feet, and if at the final pressure P_f S_w cubic feet of water has encroached into the reservoir, and S_p cubic feet of water has been produced from the reservoir, then the volume V_f after producing G_p SCF of gas is given by:

35
$$V_f = V_i - S_w + B_w S_p$$

where:

B_W = volume factor for the water in units of barrels per surface barrel.

Or the materials balance equation may be expressed in terms of moles of gas as $M_p = m_i - m_f$. Replacing the terms in the immediate above equation by their equivalence in the gas laws of Boyle and Charles, and using equation 35 above, we find that:

36
$$\frac{P_{sc}G_p}{T_{sc}} = \frac{P_1 V_1}{Z_1 T} - \frac{P_f(V_1 - S_W + B_W S_p)}{Z_f T}$$

for volumetric reservoirs where there is no water influx or water production, the preceding equation 36 reduces to the simple case of:

37
$$\frac{P_{sc}G_p}{T_{sc}} - \frac{P_1 V_1}{Z_1 T} = \frac{P_f V_1}{Z_f T}$$

where P_{sc}, T_{sc} are fixed since P_1, Z_1 and V_1 are also fixed for any particular reservoir.[14] If we make the following substitutions, letting:

$$b = \frac{P_1 V_1 T_{sc}}{Z_1 P_{sc} T}$$

and

$$m = \frac{V_1 T_{sc}}{P_{sc} T}$$

---

[14] Ibid., Chapter 12, the formulas are scattered throughout.

we get the more simplified materials balance equation in
equation 38:

38
$$G_p = b - m \frac{P_f}{Z_f}$$

which can be seen to be the equation of a straight line with
slope m.

We may again rewrite equation 36 in terms of gas
volume factors B_{gi} and B_{gf} and solve it for G_p which then
yields us:

39
$$G_p = \frac{P_i T_{sc}}{P_{sc} Z_i T} V_1 - \frac{P_f T_{sc}}{P_{sc} Z_f T} (V_1 - S_c + B_w S_p).$$

But if we substitute:

$$B_{gi} = \frac{P_i T_{sc}}{P_{sc} Z_i T} \quad SCF/ft^3$$

$$B_{gf} = \frac{P_f T_{sc}}{P_{sc} Z_f T} \quad SCF/ft^3$$

we can write equation 39 above as follows:

40
$$G_p = B_{gi} V_1 - B_{gf}(V_1 - S_c + B_w S_p).$$

But replacing V_1 by its equivalent,

$$V_1 = \frac{G}{B_{gi}}$$

we get equation 39 alone to become:

41
$$G_p = G - B_{gf} \left(\frac{G}{B_{gi}} - S + B_w S_p \right).$$

Dividing both sides of equation 41 by B_{gf} and expanding we obtain:

42
$$\frac{G_p}{B_{gf}} = G \left(\frac{1}{B_{gf}} - \frac{1}{B_{gi}} \right) + S_c - B_w S_p$$

We may express equation 42 in a more conceptual manner by saying that:

43   Production = Expansion - Water influx
                                         - Water production .

For a volumetric reservoir, the production volume equals the expansion volume and thus equation 42 becomes in this simpler case:

44
$$G_p B_{gf} = G(B_{gf} - B_{gi})$$

Equation 44 is used in water drive gas reservoirs to calculate the initial gas in place if the water influx is known, or the water influx when the initial gas in place has been determined from the known data. Where there is water influx, the initial gas in place is calculated at successive stages of depletion, and assuming no water influx, takes on successively higher values; whereas with volumetric reservoirs, the calculated values of the initial gas should remain substantially constant.

We have in the above analysis implicitly assumed that the fluid in the reservoir at all pressures as well as on the surface was in a single phase. However, this is not

normally the case. Since most gas and oil reservoirs produce some hydrocarbon liquid, commonly called condensate,
it is necessary to include this in the calculation of
initial gas or oil in place, for the condensate of a gas
reservoir may vary from a few to a hundred or more barrels
per million SCF of gas recovered. So long as the reservoir
fluid remains in a single fluid phase, the calculations
given above may be used provided that the cumulative gas
production G_p is modified to include the condensate liquid
production. However, we assume this holds in order that we
do not get entangled in the more complicated problem of
estimating liquid gas reservoirs, which would necessitate
a considerable expansion and modification of the above
results.

Oil reserves[15]

As we have seen in the previous section in

---

[15]For this section, one can see any book on pet-
roleum engineering and the calculation of oil reserves. The
terminology here used is the standard terminology recom-
mended by the API. See for example, J. C. Calhoun,
Fundamentals of Reservoir Engineering, (Norman, Okla., Uni-
versity of Oklahoma Press, 1953), Carl Gatlin, Petroleum
Engineering, Drilling and Well Completion (Englewood Cliffs,
N. J., Prentice Hall, 1960); Shell International Petroleum
Co., Inc., The Petroleum Handbook, (London, 1959); Sylvain
H. Pirson, Elements of Oil Reservoir Engineering, (New York,
McGraw Hill, 1958); and others.

calculating gas reserves, there are a number of methods to determine initial oil in place and the recovery factor. In the previous section, two methods were viewed: (1) the volumetric method and (2) the materials balance equation.

First, the volumetric method for estimating oil in place is based on log and core data to determine the bulk volume, the porosity of the oil, and fluid analysis. Under normal conditions then, one acre-foot of bulk productive rock contains:

$$\text{Interstitial water:} \quad 7758 \quad (S_w)$$
$$\text{Reservoir Oil:} \quad 7758 \quad (1\text{-}S_w)$$
$$\text{Stock tank oil:} \quad \frac{7758 \quad (1\text{-}S_w)}{B_{oi}}$$

where:

7758 barrels is equivalent to one acre-foot;

$\emptyset$ is porosity;

S_w is interstitial water as a fraction of pore volume;

B_{oi} is initial formation volume factor of the reservoir oil.

For oil reservoirs under volumetric control, since there is no water influx and hence the oil must be replaced by gas, the gas saturation increases as the oil saturation decreases. If S_g is the gas saturation factor and B_o is the oil volume factor at abandonment, then one acre-foot of rock at abandonment contains:

$$\text{Interstitial water:} \quad 7758 \ (S_w)$$
$$\text{Reservoir gas:} \quad 7758 \ (S_g)$$
$$\text{Reservoir oil:} \quad 7758 \ (1-S_w-S_g)$$
$$\text{Stock tank oil:} \quad \frac{7758 \ (1-S_w-S_g)}{B_o}$$

Then oil recovered per acre-foot is given by:

45 $$\text{Oil Recovery} = 7758 \ \emptyset \left[\frac{(1-S_w)}{B_{oi}} - \frac{(1-S_w-S_g)}{B_o} \right]$$

The factual recovery in terms of stock tank barrels is:

46 $$\text{Recovery} = 1 - \frac{(1-S_w-S_g)}{(1-S_w)} \frac{B_{oi}}{B_o}$$

For reservoirs where there is no water influx, termed volumetric reservoirs, it is possible to calculate initial oil in place using the material balance equation.[16]

When average reservoir pressure from an initial P_1 to some lower pressure P in an oil well producing above the bubble-point pressure, the initial reservoir pore volume V_{pi} declines to some value V_p owing to the formation compressibility C_f. When the average pressure drops by an

[16]Sylvain J. Pirson, Oil Reservoir Engineering, (New York, McGraw Hill, 1958), pp. 456-457.

amount dp, the pore volume at the lower pressure is:

47
$$V_p = V_{pi} (1-C \, dp)$$

As the average pressure declines, the initial connate water volume $S_w V_{pi}$ expands to a value $S_w V_{pi} (1+C_w dp)$ where C_w is the average compressibility of the reservoir water in the pressure interval $(P_i - P)$. If during the interval W_e reservoir barrels of water enter the reservoir and W_p surface barrels with volume factor B_w are produced, then the reservoir water volume at the lower pressure is:

48
$$V_w = S_w V_{pi} (1+C_w dp) + W_e - B_w W_p.$$

The difference between the two volumes $V_p - V_w$ is the volume of the undersaturated oil remaining in the reservoir at the lower pressure, or $B_o(N-N_p)$ and thus:

49
$$B_o (N-N_p) = V_{pi} (1-C_f dp) - S_w V_{pi} (1+C_w dp)$$
$$- W_e + B_w W_p$$

But the initial pore volume is $N B_{oi}/1-S_w$. Substituting this value for V_{pi} above and dividing by B_{oi}, we obtain:

50
$$N\frac{B_o}{B_{oi}} - N_p\frac{B_o}{B_{oi}} = \frac{N(1-C_f dp)}{1-S_w} - \frac{S_w N(1+C_w dp)}{1-S_w}$$
$$- \frac{W_e}{B_{oi}} + \frac{B_w W_p}{B_{oi}}$$

Rearranging terms equation 50 becomes:

51
$$N B_{oi} \left[\frac{B_o}{B_{oi}} - 1 + \frac{(S_w C_w + C_f)dp}{1-S_w} \right] = N_p B_o - W_e + B_w W_p$$

In volumetric reservoirs $W_e = 0$ and W_p is negligible so that:

52
$$N_p \, B_o = N \, B_{oi}\left[\frac{B_o}{B_{oi}} - 1 + \frac{S_w C_w + C_f)dp}{1-S_w}\right]$$

If we neglect the rock and water compressibilities that is
$C_w = 0$, $C_f = 0$, then:

$$N_p B_o = N \, B_{oi}\left[\frac{B_o}{B_{oi}} - 1\right]$$

$$N_p B_o = N \, B_o - N \, B_{oi}$$

53
$$\frac{N_p}{N} = \frac{B_o - B_{oi}}{B_o}$$

which is the initial oil in place for a reservoir that is
volumetric in nature.[17]

This equation above holds for reservoirs in which
there is no initial gas cap: Where there is a gas cap,
there is no liquid expansion of energy. However, the energy
stored in the dissolved gas is supplemented by that in the
cap, and this usually means that the oil recoveries form a
gas cap reservoir is greater. In gas cap drives, as production proceeds and reservoir pressure declines, the expansion of the gas cap displaces oil downward into the well.
At the same time, by virtue of its expansion, the gas cap

---

[17]Ibid., pp. 473-484.

retards pressure decline and, therefore, the liberation of solution gas within the oil zone, thus improving oil recovery by reducing the producing gas-oil ratio of the wells.

The second mechanism used for drive is that of water influx. Water influx into a reservoir may take the form of edge water or bottom water, the latter indicating the oil is underlain by a water zone whose movement is essentially vertical. The most important characteristics of a water-drive process are:

(1) The volume of the reservoir is constantly reduced by the water influx. This influx is a source of energy in addition to the energy of liquid expansion above the bubble point and the energy stored in the gas or gas cap.

(2) The bottom-hole pressure is related to the ratio of water influx to voidage. Where the voidage slightly exceeds the influx there is only a slight pressure decline. When the voidage is pronounced, the pressure decline is likewise very pro-nounced.

(3) For edge-water drives, regional migration
    is pronounced in the direction of the
    higher structural areas.

(4) As the water encroaches in both edge-water
    and bottom-water drives, there will be any
    increasing volume of water provided until
    only water is produced by the wells.

(5) Under favorable conditions, oil recoveries
    will range from 60% to 80% of the oil in
    place for initial recovery.

The general material balance equations, given by
Pirson,[18] simply states that since the volume of a reservoir
is constant, the algebraic sum of the volume changes in the
oil, gas, and water volumes in the reservoir must be zero.
If the assumption of equilibrium is attained at all times
in the reservoirs between the oil and its solution gas, it
is possible then to write a generalized material-balance
expression relating the quantities of oil, gas, and water
produced, the average reservoir pressure, and the quantity
of water which may have encroached from the aquifer and
finally the initial oil and gas content of the reservoir.

---

[18]Ibid., pp. 474-480.

For the derivation of the oil, gas, and water volume, it is necessary to define:

a. Changes in the oil volume:

1. Initial reservoir oil volume $= N\,B_{oi}$ ft^3

2. Oil volume at time t and pressure P
$= (N-N_p)E_o$ ft^3

3. Decrease in oil volume $= N\,B_{oi}$
$$- (N-N_p)B_o \text{ ft}^3$$

b. Changes in Gas Volume:

1. Ratio of initial free gas $= m = \dfrac{GB_{gi}}{N\,B_{oi}}$

2. Initial free gas volume $= GB_{gi} = N_m B_{oi}$

3. $\begin{bmatrix} \text{SCF free gas} \\ \text{at time t} \end{bmatrix} = \begin{bmatrix} \text{SCF initial} \\ \text{gas free} \end{bmatrix}$

$$- \begin{bmatrix} \text{SCF gas} \\ \text{produced} \end{bmatrix} - \begin{bmatrix} \text{SCF gas} \\ \text{remaining} \end{bmatrix}$$

$$G_f = \left[\frac{N_m B_{oi}}{B_{gi}} + NR_{si} \right] - \left[N_p R_p \right] - \left[N-N_p R_s \right]$$

4. Reservoir free gas volume at time t $= \left[\dfrac{N_m B_{oi}}{B_{gi}} \right.$

$$\left. + NR_{si} - N_p R_p - (N-N_p)R_s \right] B_g$$

5. Decrease in free
    volume gas
$$= N_m B_{o1} - \left[\frac{N_m B_{o1}}{B_{g1}} + \right.$$
$$NR_{s1} - N_p R_p -$$
$$\left. (N-N_p)R_s \right] B_g$$

c. Changes in water volume:

1. Initial water volume $= W$

2. Cumulative water produced at time $t = W_p$

3. Reservoir volume of cumulative produced
    water $= B_w W_p$

4. Volume of water encroached at time $t = W_e$

5. Increase in
    water volume
$$= (W + W_e - B_w W_p) - W$$
$$= W_e - B_w W_p$$

Now equating the decrease in oil volume and the
free gas volumes to the increase in water volumes given:

54
$$N B_{o1} - NB_o - N_p B_o + N_m B_{o1} - \frac{N_m B_{o1} B_g}{B_{g1}}$$
$$- NR_{s1}B_g + N_p R_p B_g + NB_g R_s - N_p B_g R_s = W_e - B_w W_p$$

Adding and subtracting $N_p B_g R_{s1}$ yields equation 55 :

55
$$N B_{o1} - N B_o + N_m B_{o1} + \frac{N_m B_{o1} B_g}{B_{g1}} - NR_{s1}B_g +$$
$$N_p R_p B_g + NB_g R_s + N_p B_g R_s - N_p B_g R_s - N_p S_g R_{s1}$$
$$= W_e - B_w W_p$$

Factoring out N and N_p we obtain:

56
$$N \, B_{oi} + N_m B_{oi} - N \quad B + (R_{si} - R_s)B_g \quad + N_p$$

$$B_o + (R_{si} - R_s)B_g \quad + (R_p - R_{si})B_g N_p \; -$$

$$\frac{N_m B_{oi} B_g}{B_{gi}} = W_e - B_w W_p$$

Letting $B_{oi} = B_{ti}$ and $(B_o + (R_{si} - R_s) B_g) = B_t$, then this equation becomes:

57
$$N(B_{ti} - B_t) + N_p \quad B_t + (R_p - R_{si})B_g \quad + N_m B_{ti}$$

$$(1 - \frac{B_g}{B_{gi}}) \; = \; W_e - B_w W_p$$

then:

58
$$N \quad B_{ti} - B_t + \frac{m B_{ti}}{B_{gi}} (B_{gi} - B_g) \quad = \; - N_p$$

$$B_t + (R_p - R_{si})B_g \quad + (W_e - B_w W_p)$$

Then N cumulative oil produced is:

59
$$N = \frac{N_p \quad B_t + (R_p - R_{si})B_g \quad - (W_e - B_w W_p)}{B_t - B_{ti} + \frac{m B_{ti}}{B_{gi}} (B_g - B_{gi})}$$

for water drive reservoir with initial gas cap. If there is no water drive, then $W_e = 0$ and then equation 59 is:

60
$$N = \frac{N_p \quad B_t + (R_p - R_{si})B_g \quad + B_w W_p}{B_t - B_{ti} + \frac{m B_{ti}}{B_{gi}} B_g - B_{gi}}$$

and where there is no free gas $m = 0$ and we find equation 59 is:

$$N = \frac{N_p \left[B_t + (R_p - R_{si})B_g \right] - (W_e - B_w W_p)}{B_t - B_{ti}}$$

If there is neither water drive nor gas cap, then the cumulative oil produced is given by:

61
$$N = \frac{N_p \left[B_t + (R_p - R_{si})B_g \right]}{B_t - B_{ti}}$$

Thus, by using equations 36 or 44 for gas or equations 59 or 61 or some variation of these equations, we can obtain a measure of the gas and/or oil reserves in a reservoir.[19] This is important to obtain the reserves since the economics of ultimate production will depend on this estimate, and thus the rate of production and well spacing will follow from this estimation. However, before considering the rate of production further, it is necessary to discuss more detail the principles involved in the flow of a fluid through a porous media.

---

[19]These equations are derived in many engineering texts; see for example, S. Pirson, Oil Reservoir Engineering, (New York, McGraw Hill, 1958), Chapter 10 for the derivation and extension of the equations into difference equation analysis.

CHAPTER 4

FLUID FLOW IN RESERVOIRS

Introduction

Up to the present, we have been concerned with
reservoir reserves and have deliberately ignored many com-
licating aspects which could have been introduced to obtain
a more realistic presentation of reservoir recovery.[1]
Having thus discussed in detail the calculation of the
reserves of a reservoir, it is next necessary in our analysis to discuss at some length the problems associated with
the flow of fluids in a reservoir in order to be able to
determine the rate of production of a reservoir. In the
analysis of the flow of fluids, our purpose is to include
enough details so that the reader will be aware of the
engineering phenomena we are describing, and secondly, to

---

[1]J. C. Calhoun, Fundamentals of Reservoir
Engineering, (Norman, Okla., The University of Oklahoma
Press, 1953), Part II, pp. 75-196; or S. J. Pirson, Oil
Reservoir Engineering, (New York, McGraw-Hill, 1958)
Chapter 8, pp. 389-440, and others.

understand in some detail the variables which determine
fluid flow.

In Chapter 3, we defined the functional relation
of fluid flows as follows:

$$F = F\,[\text{viscosity, compressibility, flows system,}\\ \text{pressure build-up, drainage radius}]\,.$$

In this section, we will concentrate on understanding these
variables in order that we might determine production for
both gas and oil. Likewise, we shall, in discussing fluids,
have to be concerned with the problem of steady or unsteady
flow, pressure behavior under either conditions of flow
and other important problems surrounding this area of flow
analysis.

Before attempting to delve into an analysis of
fluids in a reservoir, it is necessary to lay the foundation
with an analysis of Darcy's Law, the backbone of the study
of fluids. Darcy's Law states that: "The velocity of a
homogeneous fluid in a porous medium is proportional to the
pressure gradient and inversely proportional to the fluid
viscosity

62
$$V = \frac{k}{\mu}\frac{dP}{dS}$$

where:

V = apparent velocity in centimeters per second;

q = volumetric flow in cm^3/sec;

A = apparent total cross sectional area of rock in cm^2;

k = permeability of rock in either Darcy's or millidarcy's;

μ = fluid viscosity in centipoise;

$\frac{dP}{dS}$ = atmospheres per centimeter."[2]

Darcy's Law applies only in the region of stream-line flow; for in turbulent flows, which occur at higher velocities, the pressure gradient increases at a faster rate than the flow rate. In most cases, except for those instances of very large production rates or very large infection rates near the bore, we can make the assumption of streamline flow without much loss to reality and thus, assume that Darcy's Law does hold. The law itself does not apply to the flow within a particular individual pore channel, but rather only to a portion of the rock, that is, it is a statistical law which averages the behavior of many pore channels instead of treating each channel separately.

Owing to the porosity of the rock and the tortusity[3] of the flow paths, and the absence of flow in

[2]Donald Katz, et. al., Handbook of Natural Gas Engineering, (New York, McGraw-Hill, 1959), p. 40.

[3]By tortusity we mean twisted in form or crooked in shape.

some of the pore spaces, the actual fluid velocity may vary from point to point in the rock, and thus, maintain an apparent average velocity which we then term the velocity of the fluid. Because actual velocities are usually not measurable, and to keep permeability and porosity separated, the concept of apparent velocity is the underlining basis of Darcy's Law. This means that the actual forward velocity of a fluid will be the apparent velocity divided by the porosity of the substance.

The unit of permeability is called the "darcy." A rock of one darcy permeability is: "One in which fluid of one centipoise viscosity will move at a velocity of one centimeter per second under a pressure gradient of one atmosphere per centimeter."[4] Since this is a fairly large unit for most producing rock, permeability is usually measured in units one thousandth as large, commonly called "millidarcy" (MD).

The gradient dP/dS is the driving force, and in our analysis, we assume that it is due to the fluid pressure gradient only. We shall ignore hydraulic gradients because of the complications they entail as well as the

------

[4]See Pirson, Oil Reservoir Engineering, (New York, McGraw-Hill, 1958), p. 65.

fact that we are only concerned with reservoirs under natural drive.

Viscosity

The viscosity of natural gas and natural gas liquid depends on the temperature, the pressure, and the physical composition of the gas. Viscosity can be defined by the relationship:

63
$$W = \frac{\mu}{g_c} \frac{d\mu}{dy}$$

where:

$\mu =$ coefficient of viscosity in grams mass/cm.sec;

$W =$ shear stress per unit area in the shear plan parallel to the direction of the flow;

$g_c =$ conversion factor in grams mass/gram force/ cm.sec^2;

$\frac{d\mu}{dy} =$ velocity gradient perpendicular to the plane of shear in cm/cm sec^2.

The importance of viscosity in calculating fluid flow will not be seen until we study Parseuilles Law of capillary flow. We shall detain considerations of this problem until we have further developed some of the necessary concepts of fluid flow.

The viscosity for oil depends on whether the oil in the reservoir is above a below bubble point pressure.[5]

---

[5]*Ibid.*, p. 342 for gas.

When it is below the bubble point, the viscosity decreases with increasing pressure because of the thinning effect of gas entering the solution. However, if pressure is above bubble point, then the viscosity increases with increasing pressure.[6]

Compressibility

For most engineering problems, the fluid in a reservoir may be classed as either: (1) incompressible; (2) compressible liquid; or (3) gas. We are obviously interested in the nature of the fluid because the nature of the fluid enters not only the calculation of viscosity, but also is important in the determination of the compress-ibility. We shall be concerned with all three natures, that is, the compressibility of gas, incompressible and compress-ible liquid, and shall devote this section to considerations of compressibility.

In order to arrive at the concept of gas compress-ibility, we need to differentiate the real gas law given by:

64
$$V = \frac{Z \mu R T}{P}$$

where:

V = velocity;

[6]Ibid., pp. 347-351 for oil reservoirs.

Z = compressibility factor;

R = universal gas constant;

T = temperature;

P = pressure;

differentiating with respect to pressure under the assumption of equal temperature and a universal gas constant, we arrive at:

65
$$\frac{dV}{dP} = \frac{\mu RT}{P} \cdot \frac{dZ}{dP} - \frac{Z\mu RT}{P^2}$$

multiplying and dividing the first term by Z and the second term factoring out $1/P$, we obtain:

66
$$\frac{dV}{dP} = \frac{Z\mu RT}{P} \; \frac{1}{Z} \frac{dZ}{dP} - \left[\frac{Z\mu RT}{P}\right] \frac{1}{P}.$$

substituting;

67
$$\frac{1}{V} \frac{dV}{dP} = \frac{1}{Z} \frac{dZ}{dP} - \frac{1}{P}.$$

But defining

68
$$c = -\frac{1}{V} \frac{dV}{dP}$$

and

69
$$C_g = \frac{1}{P} - \frac{1}{Z} \frac{dZ}{dP}$$

and if we allow the gas factor to be 1.00, i.e., an ideal gas, and $\frac{dZ}{dP} = 0$, then C_g, i.e., gas compressibility, is merely $1/P$.[7]

In the study of transient flows in reservoirs, the diversibility constant $k/\mu c\emptyset$ enters the equation where:

k = absolute permeability;

μ = viscosity;

c = compressibility;

$\emptyset$ = porosity;

If there is only one fluid in the reservoir, rock compressibility is neglected and then the compressibility is merely the compressibility of the fluid, and the porosity is merely the effective porosity. We shall, in our analysis, assume this to be the case and thus shall exclude the phase phenomena and the complications which arise therein.

A compressible liquid can be defined as one whose volume changes with pressure and can be expressed by the equation:

70
$$V = V_1 \, e^{\, c(P_1 - P)}.$$

Letting $e^{\, c(P_1 - P)} = e^X$, we can represent e^X by a series expansion:

71
$$e^X = 1 + X + \frac{X^2}{2!} + \frac{X^3}{3!} + \ldots \frac{X^n}{n!}.$$

---

[7]See D. Katz, et. al., Handbook of Gas Engineering (New York, McGraw-Hill, 1958), p. 501.

Then:

72
$$V = V_1 \left[1 + C(P_1 - P) \right]$$

which is the volume charge for a compressible liquid.

Needless to say, for the incompressible liquid, that is, a liquid whose volume does not change with pressure, the volume equation is sufficient for its consideration in the flow process.

Flow Systems

There are a number of ways in which to classify flow reservoir systems. Here we shall adapt the most common system described by Muscat, Frick, Uren and others. In their works, they classify reservoir systems according to: (a) the type of fluid; (b) the geometry of the reservoir; (c) the rate at which the flow approaches a steady state equilibrium following a disturbance in the reservoir.

Fluids are typed according to their compressibility, that is, repeating what was described in the previous section either as incompressible, compressible, or gas. In addition, they can be classified by either single-phase flow, two-phase flow or at most, three-phase flow. Most systems are single-phase flow, that is, either just gas, oil or water, and we shall be concerned only with the

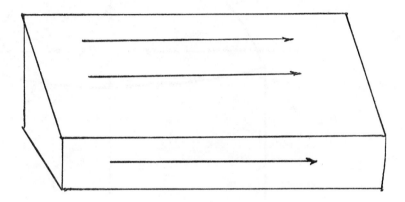

Figure 10   Linear Flow System

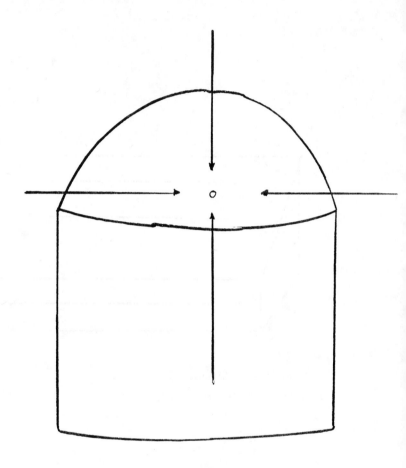

Figure 11   Radical Flow System

single-phase flow of gas or oil. An example of two-phase flow would be that of gas and oil, or gas and water, or oil and water. The three-phase flow is obviously gas, oil and water.

The types of geometries that are of interest to us are either the linear or the radical flow or a combination of linear-radical flow. A linear flow is described in Figure 10, the flow lines being parallel as shown and the cross-section exposed to flow being constant.

In the case of radical flow, as is shown in Figure 11, the flow lines are straight and converge to a common center, say the center of a well bore. In actuality, neither of these flow patterns or geometries are found in petroleum reservoirs; however, they do yield results good enough that reservoirs can be approximated by these idealized forms.

Finally, a flow system in a reservoir can be classed as either a steady-state flow or unsteady-state flow. In a steady-state system, the pressure and the fluid, that is, the fluid velocity at every point throughout the system, adjust instantaneously to a change in pressure or flow rate of any part of the system. No real life system responds that quickly to the response of a change, but the

point is that the readjustment period is very small. For
a radical flow system, such as would be encountered in the
flow toward a producing well bore, the area A of the flow
path is a variable in the radical flow that must be taken
into consideration in the integration of Darcy's Law:

73
$$V = \frac{q}{A} = -\frac{k}{\mu}\frac{dP}{dS}$$

Letting A = 2πrh where r is radius in feet and h is the
formation thickness in feet, then:

74
$$\frac{.0004476\ \pi hkP\ (\pm dP)}{ZT\mu\ \frac{dS}{s}}$$

Integrating the radical flow equation for constant values of
k, T, h, μ, and Z, we obtain:

75
$$V = \frac{\pm.0007030\ hk\ (P_1 - P_2)}{ZT\mu\ln \frac{r_1}{r_2}}$$

where we define:

h = as sand thickness in feet;

k = permeability in millidarcys;

P = pressure;

Z = compressibility;

T = flowing temperature in the well;

μ = viscosity;

$V =$ gas flow rate measured at 14.7 psia.[8]

This will now be referred to as the field units for gas

flow. The term $\ln (r_1/r_2)$ is the natural logarithm of the

ratio of the radii measured from the center of the well

bore.

Linear flow for fluids in steady state, assuming

first that the ends only are open for flow, can be given

by:

76
$$V = \frac{q}{A} = -1.127 \ \frac{k}{\mu} \ \frac{dP}{dX}$$

Integrating over the length of the porous body:

77
$$\frac{q}{A} \int_0^L dx = -1.127 \ \frac{k}{\mu} \int_{P_1}^{P_2} dP$$

yields:

78
$$q = \frac{1.127 \ kA \ (P_1 - P_2)}{\mu L}$$

For radical flow of an incompressible fluid in

steady state, assuming uniform thickness and permeability,

the flow across any circumference is a constant. If we

assume P_w is the well bore pressure when the well is flowing

q barrels per day, and pressure P_e is maintained and the

---

[8]Note that the sign for dp is - for injection and
+ for the production from the reservoir.

external well radius is r_e, then at this radius r_e:

79
$$V = \frac{q}{A} = \frac{q}{2\,rh} = -1.127\ \frac{k}{\mu}\frac{dP}{dr}$$

Integrating between any two radii r_1 and r_2 for pressures P_1 and P_2, we obtain:

80
$$\int_{r_1}^{r_2} \frac{q}{2\,rh}\ dr = -\int_{P_1}^{P_2} 1.127\ \frac{k}{\mu}\ dP$$

and q is then:

81
$$q = \frac{708\ kh\ (P_2 - P_1)}{\mu\ \ln\ (r_1/r_2)}$$

If the two radii integrated between are the well bore radius r_w and the drainage radius r_e, then we obtain for an incompressible fluid:

82
$$q_{sc} = \frac{708\ kh\ (P_e - P_w)}{\mu\ B_o\ \ln\ (r_e/r_w)}$$

Brounscombe and Collins[9] have derived an equation similar to the above for a steady state compressible liquid flow in reservoirs. The equation they derive is given by:

83
$$q_{sc} = \frac{708\ kh\ (P_e - P_w)}{\mu\ B_o\ \left[\ln\ r_e/r_w - \frac{1}{2}\right]}$$

---

[9]E. R. Brounscombe and F. Collins, "Pressure Distribution in Unsaturated Oil Reservoirs," Trans. AIME, (1950), Vol. 189, p. 371.

The ratio of the rate of production in an oil well given
above to the pressure drawdown at the midpoint of the producing interval is called the productivity index J, and is
given by:

84
$$ J = \frac{q_{sc}}{(P_e - P_w)} \quad \text{in} \quad \frac{bbls}{day\ psi} \cdot $$

It measures the ability of the well to produce and the maximum well flow rate possible. As the well is depleted, the
productivity index declines as the oil viscosity increases
and as the gas is released from solution, and also as the
oil looses permeability.

Thus, the maximum rate at which a well can produce
depends on the productivity index under current reservoir
conditions and on the pressure drawdown.[10]

Parseuilles Law[11]

Although the pore space within a rock is never
a straight line, or a smoothed well surface of constant
diameter, we need to make these assumptions in order to
treat capillary action in reservoir rock. Consider a

---

[10]S. Pirson, Oil Reservoir Engineering, (New York,
McGraw-Hill, 1958), pp. 656-657.

[11]Ibid., pp. 98-99. For a discussion of
Parseuilles Law as well as many other petroleum engineering
text.

capillary tube of length L in centimeters, and assume that
the radius of the tube is r in centimeters, in which is
flowing a fluid or gas of μ poises viscosity under a pressure difference of $(P_1 - P_2)$ dynes per square centimeter.
If the fluid wets the walled surface, and this is not the
case in gas save for the natural gas liquid, then the
velocity at the edge of the walls is zero and rises to a
maximum at the center of the capillary. The flow, which
may be viewed as a series of concentric shells moving at
various velocities and thus exerting viscous forces on each
other, is expressed by the relation:

85 $$F = \mu A \frac{dV}{dX}$$

where:

F = the force in dynes;

μ = viscosity in poises;

A = area in square centimeters;

$\frac{dV}{dX}$ = velocity gradient in cm/sec/cm.

The volume rate through an element dr in thickness is given
by:

86 $$dq = vdA$$

where:

dA = $2\pi rdr$.

Then the total flow rate through the capillary at these

various individual elemental flows is:

87
$$q = \int_o^q d_q = \int_o^{r_o} VdA$$

88
$$q = \int_o^{r_o} \frac{(P_1 - P_2)(r_o^2 - r^2)}{4\mu L} 2\pi rdr.$$

Integrating we obtain:

89
$$q = \frac{\pi r_o^4 (P_1 - P_2)}{8\mu L}$$

which is called Parseuille's Law of viscous flow of liquids

through capillaries. The same above equation has been

derived by Klinkerberg for gases.[12] A common sense compari-

son of the two leads to the prediction that the permeability

to gases will be higher than to liquids, and the differences

will be greater for lower permeable rocks, lower gas flow

pressures, and lower molecular weight gases.[13]

Many times in gas and oil reservoirs, the flow is

interrupted by a fracture, or is dependent on the flow

through a fracture. Assuming a constant width, the flow

through this fracture may be found by applying the formula:

---

[12]L. J. Klinkerberg, "Permeability of Porous Media
to Liquids and Gases," Drilling and Production Practices
API, (1941), p. 200.

[13]Ibid., p. 200

90
$$q = \frac{W^2 A (P_1 - P_2)}{12 \mu L}$$

where:

W = width of fracture in centimeters;

A = cross-sectional area of the fracture;

(P_1-P_2) = pressure differential which exists at the ends of the fracture;

μ = viscosity;

L = the length of fracture in centimeters.

This phenomenon of fracture is very important in the cases of limestone and sandstone rock. Fractures in many cases account for the economic production rates which make the well profitable as a result of these openings.

Production

We assume a gas well is producing q_{sc} SCF/day under standard conditions per day in a steady state radical flow. The volumetric flow rate q at a given radius r under pressure P will be given by:

91
$$q = \frac{P_{sc} q_{sc} TZ}{5.615 \, T_{sc} P}$$

But since $q/A = -1.127 \, k/\mu \, dp/dr$ from Darcy's equation, and since A = 2 rh, we obtain:

92
$$-1.127 \, \frac{k}{\mu} \frac{dP}{dr} = \frac{P_{sc} q_{sc} TZ}{5.615 \, T_{sc} \, P2\pi rh}$$

Integrating equation 92 between P_w and P_e and r_w and r_e, we obtain:

93
$$q_{sc} = \frac{19.88 \; T_{sc}hk \; (P_e^2 - P_w^2)}{P_{sc} \; TZ \, \mu \ln (r_e/r_w)}$$

If we let $T_{sc} = 520°$ R and $P_{sc} = 14.7$ psia, then we obtain:

94
$$q_{sc} = \frac{703 \; kh \; (P_e^2 - P_w^2)}{TZ \; \ln (r_e/r_w)}$$

where we define everything as before and we make the assumption that the flow across the external boundary is equal to that produced at the well. These equations can then be used to find the average formation permeability for gas flow. They also form the basic equations for gas well testing, i.e., flow testing. However, turbulence near the well bore and unsteady state factors may cause behavior which is unaccountable for by or from the above equations.

Formerly, when gas was not the important commodity it is today, it was customary to rate and test gas wells on the basis of their open-flow capacity, which, of course, was obtained by allowing the well to flow freely into the atmosphere. However, not to prevent waste and possible formation damage, the open flow capacity for a gas well is now obtained by extrapolating tests results made at several intervals rather than by allowing the well to flow free at

full rate. Since controlling, or back-pressure, is maintained at the surface, these tests are commonly called back-
pressure tests.[13] In order to eliminate the differences in
flow rates due to the length and size of the pipe through
which the gas flows, the bottom-hole pressure and the open-
flow capacity of the potential of a gas well is defined as
the rate of production, in SCF/day, the well would have
produced if its flowing bottom-hole pressure were reduced to
atmospheric pressure. This is a rather theoretical figure
and it is obtained by extrapolating a log-log plot of
$(P_e^2 - P_w^2)$ versus q_{sc} to $P_w = P_{atm}$, assuming μ, T and Z are
constant and that a linear flow exists. Taking the logs of
equation 94, we obtain equation

95 $$\log q_{sc} = \log \left[\frac{703\ kh}{\mu TZ \ln r_e/r_w} \right] + \log (P_e^2 - P_w^2)$$

which says that for an ideal gas the log-log plot of q_{sc}
versus $(P_e^2 - P_w^2)$ is a straight line with slope of 45°. In
practical engineering analysis, most gas wells do yield data
which are linear for the log-log plot with slopes very near
to 1.000.[14] In others, due to turbulence and other factors,

---

[13]See for example, D. Katz, et. al., Handbook of
Gas Engineering, (New York, McGraw-Hill, 1958), p. 502.

[14]Ibid., p. 582 for log-log plots for various
fields and reservoirs studied by the authors.

the plots may be non-linear and their slopes less than one.
The above equations 94 have been modified to include these
wells as follows:

$$96 \qquad q_{sc} = \frac{703 \text{ kh } (P_e^2 - P_w^2)}{\mu \text{ TZ } \ln (r_e/r_w)}$$

But if we let $C = 703 \text{ kh}/ \text{ TZ } \ln (r_e/r_w)$, we obtain:

$$97 \qquad q_{sc} = C (P_e^2 - P_w^2)$$

But taking the logs of both sides yields us:

$$98 \qquad \log q_{sc} = \log C - \log (P_e^2 - P_w^2)$$

In the case of attempting to measure the open-flow
potential or back-pressure testing, the flow at any given
rate should be continued long enough for the reservoir to
approach a steady-state condition. The important question
is the time necessary to reach this steady-state condition,
and this can be estimated by use of the readjustment time
equation given by:

$$99 \qquad T_r = \frac{\Delta V}{q_{sc} B_o} = \frac{.04 \, \mu C \phi r_e^2}{k}$$

McRoberts[15] has also derived an approximate
equation for the variation in drainage radius with time for

[15]D. J. McRoberts, "Effects of Transient Conditions
in Gas Reservoir," Trans. AIME, (1949), Vol. 186, p. 36.

gas well flows, assuming a logarithmic pressure distribution
is established between the well and the drainage radius at
any time, then the readjustment time is given by:

100
$$T_D = \frac{.08 \mu \emptyset r_e^2}{k \ (P_e + P_w)} \left[1 - \frac{1}{2 \ln \ (r_e/r_w)} \right]$$

For example, assume a gas well has a radius of $r_w = .333$ ft.
and is shut-in long enough for the pressure in the well to
reach the same everywhere, $P_e = 1000$ psia. If the gas
porosity is 20%, the viscosity .015 cp, and the permeability
is 100 millidarcys, with a bottom-hole pressure of 800 psia
the time for the drainage radius to reach 1,320 ft. (or an
80 acre spacing radius) will be:

101
$$T_D = \frac{.08 \ x \ .015 \ x \ .20 \ x \ 1320^2}{.100 \ (1000 - 800)}$$
$$\left[1 - \frac{1}{2 \ln} \quad \frac{1326}{.333} \right]$$

$$T_D = 2.18 \text{ days}$$

In another article of the Trans. AIME in 1955, Cullender[16]
found that the variable results of many gas well tests are
due to the complex nature of the pressure disturbances which
are created about the wells as the result of the tests

_____

[16]M. A. Cullender, "The Isochronal Performance
Method of Determining the Flow Characteristics of Gas Well,"
Trans. AIME, (1955), Vol. 204, p. 137.

conducted on the well. He thus proposed in his article that a long enough period for test flow be allowed so that each flow would begin with the same pressure distribution in the reservoir. He also concluded that as a result of equation 100 the external radius increased with time, and the effect on equation 97 was to cause the value of C to decrease over time. At the same time, after starting flow at any selected rate from the same set of initial conditions, the drainage radius should be the same, and thus C likewise. Since the exponent n is not connected with the increasing drainage radius, the curve plotted on a log-log paper for the various times should have the same slope. This is called by Cullender the isochronal performance method of testing gas wells.

The following figures 12, −18, show some iso-chronal performance curves for a gas well studied by Cullender. A slope of .948 was obtained by him regardless of the time of flow. A shift of the curves to the left indicates a decline in the constant C owing to an increase in the drainage radius with time. The other figures show the value of C for the same well plotted against time on semi-log paper. It appears to be approaching a constant value near 40.

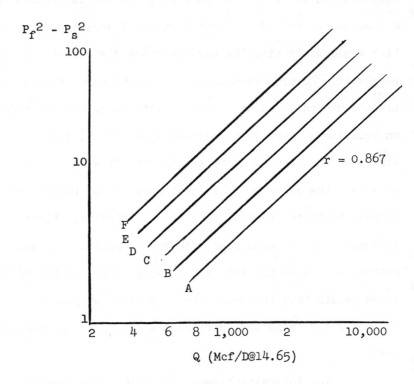

$P_f^2 - P_s^2$

Q (Mcf/D@14.65)

Figure 12   Isochronal Performance Curves of Gas
            Well No. 1 (Slope = 0.867)

Note: Curve A;0.1-hour duration of flow.
            Curve B;0.2-hour duration of flow.
            Curve C;0.5-hour duration of flow.
            Curve D;1.0-hour duration of flow.
            Curve E;3.0-hour duration of flow.
            Curve F;24.0-hour duration of flow.

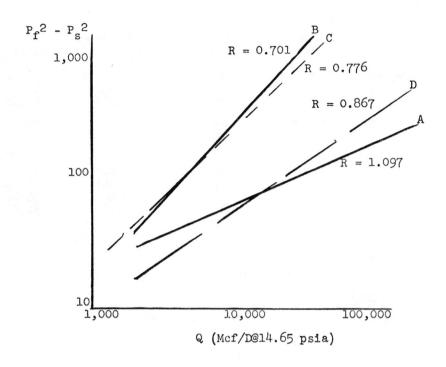

Figure 13   Back-Pressure Test Data of
Gas Well No. 1

Note: Curve A; 24-hour, reverse sequence
             back-pressure test (slope = 1.097).
             Curve B; 24-hour, normal sequence
             back-pressure test (slope = 0.701).
             Curve C; 24-hour, normal sequence
             back-pressure test (slope = 0.776).
             Curve D; 24-hour, isochronal performance
             curve (slope = 0.867).

Table 2.   Back-Pressure Test Data of Gas Well No. 1

Date	SIP, psia	Duration of Flow, Hours	Mcf/D @ 14.65 psia	$P_f^2-P_s^2$
10-3-44	435.2	24	9,900	97.70
		24	7,091	70.73
		24	4,360	46.16
10-24-44	423.6	23	4,440	38.67
		25	6,982	75.17
		22	8,212	92.35
12-11-45	394.7	24	1,947	14.56
		24	2,841	25.07
		26	3,941	38.82
		22	5,165	50.53

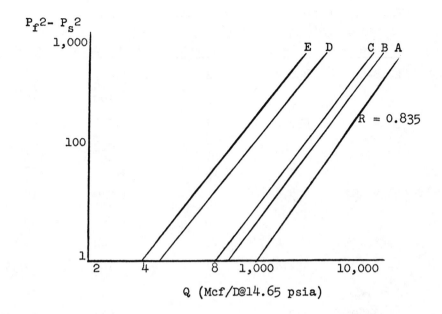

$P_f^2 - P_s^2$

Q (Mcf/D@14.65 psia)

Figure 14 Isochronal Performance Curves of Gas
Well No. 2 (n=0.835) (Slope=0.835)

Note: Curve A; 1-hour duration of flow.
            Curve B; 2-hour duration of flow.
            Curve C; 3-hour duration of flow.
            Curve D;24-hour duration of flow.
            Curve E;72-hour duration of flow.

Table 3. Performance Data of Gas Well No. 2

Date	SIP, psia	Duration of Flow, Hours	Mcf/D @ 14.65 psia	$P_f^2 - P_s^2$
2-18-48	436.0	1	1,224	8.85
		2	1,215	11.70
		3	1,200	16.77
3-2-48	435.8	1	4,262	52.22
		2	4,114	63.35
		3	4,022	70.28
		24	3,495	106.70
		72	3,238	122.43
3-23-48	434.6	1	1,710	17.09
		2	1,691	21.11
		3	1,680	23.77
		24	1,599	40.46
		72	1,562	49.57
7-25-49	434.4	1	2,107	22.77
		2	2,073	28.34
		3	2,054	31.88
7-26-49	432.7	1	3,057	35.20
		2	2,986	43.09
		3	2,942	48.07
7-27-49	432.4	1	4,208	51.39
		2	4,061	62.59
		3	3,963	69.23

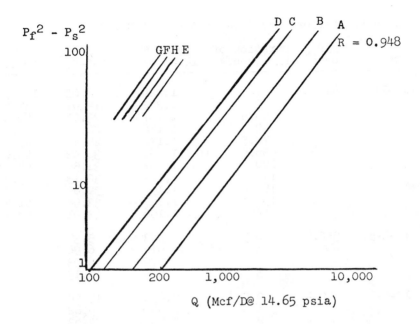

$$P_f^2 - P_s^2$$

Q (Mcf/D@ 14.65 psia)

Figure 15 Isochronal Performance Curves of Gas
Well No. 3 (Slope = 0.948)

Note: Curve A; 0.5-hour duration of flow.
           Curve B; 1.0-hour duration of flow.
           Curve C; 2.0-hour duration of flow.
           Curve D; 3.0-hour duration of flow.
           Curve E;23.5-hour duration of flow.
           Curve F;70.5-hour duration of flow.
           Curve G; 9-day duration of flow.
           Curve H; average 3-day performance.

Table 4.   Performance Data of Gas Well No. 3

Date	SIP, psia	Duration of Flow, Hours	Mcf/D @ 14.65 psia	$P_f^2 - P_s^2$	Coef.
10-11-44	441.6	1	1,229	8.62	159.5
		9	1,202	18.01	77.6
		23.5	1,187	23.07	60.6
		49	1,176	26.71	52.2
		70.5	1,171	28.81	48.4
		96.5	1,166	30.52	45.6
		120	1,163	31.56	44.1
		144	1,161	32.13	43.3
		169	1,159	32.89	42.3
		190	1,157	33.54	41.4
		214	1,156	33.91	40.9
12-3-51	352.4	0.5	983	5.37	
		1	977	6.96	
		2	970	8.93	
		3	965	10.19	
12-4-51	352.3	0.5	2,631	15.54	
		1	2,588	19.82	
		2	2,533	24.63	
		3	2,500	27.72	
12-5-51	351.0	0.5	3,654	21.63	
		1	3,565	27.40	
		2	3,453	34.03	
		3	3,390	37.97	
12-6-51	349.5	0.5	4,782	28.84	
		1	4,625	35.96	
		2	4,438	43.98	
		3	4,318	48.96	

Annual Three-Day Production Test Data

Date	SIP, psia	Duration of Flow, Hours	Mcf/D @ 14.65 psia	$P_f^2 - P_s^2$	Coef.
4-5-45	417.7	72.5	1,818	43.94	
7-13-45	403.2	72	1,848	41.16	
5-10-46	389.7	72	1,665	35.57	
5-28-48	389.9	72.25	1,457	34.65	
5-17-48	378.5	71.83	1,269	28.81	
6-17-49	371.7	72	1,389	31.96	
5-23-50	365.5	71.75	1,438	34.16	
4-29-51	355.0	72.25	1,195	27.23	
5-26-52	348.5	71.5	1.073	24.97	
7-7-53	336.1	72	1,164	27.43	

125

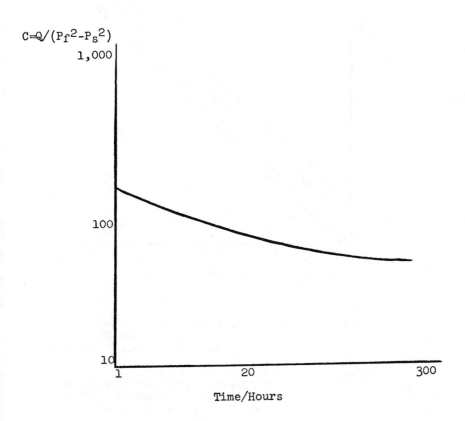

Figure 16   Relationship of Coefficient of Performance and Time of Gas Well No. 3

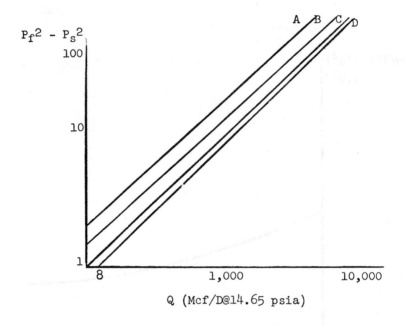

$P_f^2 - P_s^2$

Q (Mcf/D@14.65 psia)

Figure 17   Isochronal Performance Curves of Gas
Well No. 4:   Below 800 Mcf/D (Slope
= 1.000)   Above 800 Mcf/D (Slope = 0.859)

Note: Curve A; 0.5-hour duration of flow.
               Curve B; 1.0-hour duration of flow.
               Curve C; 2.0-hour duration of flow.
               Curve D; 3.0-hour duration of flow.

Table 5. Performance Data of Gas Well No. 4

Date	SIP, psia	Duration of Flow, Hours	Mcf/D @ 14.65 psia	$P_f^2 - P_s^2$
9-28-49	449.9	0.5	76	0.95
		1	76	1.16
		2	76	1.26
		3	76	1.47
9-30-49	450.2	0.5	143	1.79
		1	143	2.00
		2	143	2.42
		3	143	2.61
10-3-49	449.6	0.5	576	6.62
		1	575	8.06
		2	573	9.72
		3	572	10.94
10-4-49	449.6	0.5	1,237	16.96
		1	1,229	19.28
		2	1,219	22.49
		3	1,212	24.48
10-5-49	449.3	0.5	2,148	29.33
		1	2,116	35.37
		2	2,083	41.04
		3	2,065	44.56
10-6-49	448.5	0.5	3,079	48.22
		1	3,017	55.08
		2	2,947	63.52
		3	2,902	68.25
10-7-49	447.4	0.5	4,218	65.70
		1	4,087	76.22
		2	3,936	86.85
		3	3,852	92.87
10-11-49	448.6	0.5	5,373	86.49
		1	5,136	99.42
		2	4,887	111.71
		3	4,756	118.45
10-14-49	448.7	0.5	326	3.83
		1	325	4.67
		2	324	5.50
		3	323	6.12

Table 5. (cont'd.)

Date	SIP, psia	Duration of Flow, Hours	Mcf/D @ 14.65 psia	$P_f^2-P_s^2$
11-2-49	449.3	0.5	888	10.03
		1	881	12.49
		2	874	15.14
		3	870	16.86

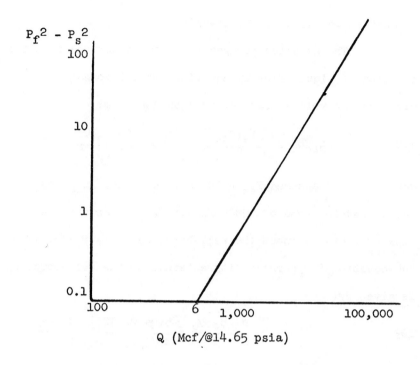

Figure 18 Stabilized Performance Curve of Gas
Well No. 5 (Slope = 0.554).

Table 6. Performance Data of Gas Well No. 5

Date	SIP, psia	Duration of Flow, Hours	Mcf/D @ 14.65 psia	$P_f^2-P_s^2$
3-30-50	439.0	1	8,373	11.27
		1	12,484	23.25
		1	16,817	40.60
4-3-50	439.9	1	570	0.09
4-4-50	439.6	1	2,231	1.01
4-5-50	439.8	1	4,841	4.32

Production and Pressure Build-up[17]

The equation for gas flow may be derived from the equation of liquid flow by equating the oil voidage rate with the gas voidage rate or in equation form:

102
$$q_{sc}B_o = \frac{P_{sc}q_{sc}ZT_n}{5.615\,T_{sc}P_{avg}} = \frac{5.04\,Q_{sc}TZ}{P_{avg}}$$

where we have assumed $T_{sc} = 60°$ F., $P_{sc} = 14.6_{psia}$ and Q_{sc} is defined in terms of MMSCF/day. P_{avg} is an average flowing pressure about the well-bore, near $(P_f + P_e)/2$. Then the equation for permeability external to the well-bore, k_e, is given by:

103
$$k_e = \frac{q_{sc}B_o\,\mu_o}{6.15mh} = \frac{5.04\,Q_{sc}TZ}{P_{avg}}\left[\frac{\mu\,g}{6.15mh}\right]$$

Then the productivity ratio is the ratio of k_{avg} given by equation 94 to k_e given by equation 103. Writing $q_{sc} = 1000\,Q_{sc}$, where Q_{sc} is in MMSCF, we obtain the productivity ratio as:

$$P.R. = \frac{100\,Q_{sc}\,\mu_g T\,\ln(r_e/r_w)}{703\,h\,(P_e^2 - P_w^2)} \quad \frac{mP_{avg}h}{.819\,Q_{sc}T\,\mu_g}$$

104
$$P.R. = \frac{4 \cdot mP\,\ln(r_e/r_w)}{(P_e^2 - P_w^2)}$$

---

[17]See S. Pirson, Oil Reservoir Engineering, (New York, McGraw-Hill, 1958) pp. 422-426.

Production, Unsteady State

Following a period of unsteady-state flow, a condition is reached in the reservoir where the pressure at every point is dropping at nearly the same rate. Neglecting small differences in the gas deviation factor, since no gas crosses the external radius r_e, the gas flow in SCF/day across any radius r must be proportional to the gas pore volume from that radius r out to r_e or in equation form:

105
$$\frac{g_r}{g_w} = \frac{\pi r_e^2 \, h\phi - \pi r^2 \, h\phi}{\pi \, r_e^2 \, h\phi}$$

then:

106
$$g_r = g_w \, (1 - r^2/r_e^2)$$

If this equation is now substituted into equation 93 in place of the constant value of q at all radii, then we obtain equation 107:

107
$$- 1.127 \, \frac{k \, dP}{\mu \, dr} = \frac{P_{sc} TZq_w \, (1-r^2/r_e^2)}{5.615 \, T_{sc} \, P \pi r h}$$

Integrating between the limits r_e and r_w and P_e and P_w, we obtain:

108
$$\int_{r_w}^{r_e} (1-r^2/r_e^2) \, \frac{dr}{r} = \int_{r_w}^{r_e} \frac{dr}{r} - \int_{r_w}^{r_e} r\frac{dr}{r_e^2}$$

which is equal to:

109
$$\frac{- 1.127 \ T_{sc} \ 5.615 \ (2 \ \pi rh)}{q_w \ \mathcal{M} \ P_{sc} \ TZ} \int_{r_w}^{r_e} P dP$$

Letting $T_{sc} = 520$ R and $P_{sc} = 14.7$ psia, we obtain qw as:

110
$$q_w = \frac{703 \ kh \ (P_e^{\ 2} - P_w^{\ 2})}{TZ \ \ln \ r_e/r_w - \frac{1}{2}}$$

But we know that $- \frac{1}{2} = \ln 0.61$ and then $\ln r_e/r_w - \ln 0.61$

$- \ln (0.61 \ r_e/r_w)$ so that qw in equation 110 then becomes:

111
$$q_w = \frac{703 \ kh \ (P_e 2 - P_w 2)}{\mathcal{M} TZ \ln \ (0.61) \ r_e/r_w}$$

The average pressure, P_{avg}, of the gas in the
drainage area may be found by either: (a) measurement with
a subsurface pressure gage when the well has been shut off
for a sufficiently long period of time; (b) or by use of the
materials balance equation. As the average pressure P_{avg} is
lower than the external pressure P_e, if it replaces P_e in
the equation 111, then the denominator must be lowered to
preserve the equality. Further, Aronofsky and Jenkins[18]
have shown that we may replace the constant 0.61 by 0.472
so that we arrive at the standard equation for gas flow
which is given below:

_____

[18]J. S. Aronofsky and R. Jenkins, "A Simplified
Analysis of Unsteady Radical for Flow," Trans. AIME, (1954),
Vol. 201, p. 149.

$$112 \qquad q_{sc} = \frac{703 \ kh \ (P_{avg}^2 - P_w^2)}{\mu TZ \ ln \ .472 \ (r_e/r_w)}$$

Equation 112 above may be combined with the material-balance equation (or the volumetric equation) to give an equation relating the fractional recovery with the rate of production, the radius of the reservoir unit, and the minimum pressure at which the well will flow. Thus, the initial gas in place for a uniformly circular reservoir at initial pressure P_1 which contains an ideal gas = 1.000 can be given by:

$$113 \qquad G_p = \pi r_e^2 \ h\phi_{hc} \ \frac{P_1}{P_{sc}} \ \frac{T_{sc}}{T_r}$$

measured in SCF. If the well has been producing at a constant rate of q_{sc} SCF/day for a period of t days, then the gas remaining at the end of the tth day and remaining to be produced is:

$$114 \qquad G_r = (G_p - q_{sc}t)$$

For an ideal gas, the gas remaining at any time will be proportional to the average pressure and thus:

$$\frac{P_1}{G_p} = \frac{P_{avg}}{G_p - q_{sc}t}$$

$$115 \qquad P_{avg} = (1 - \frac{q_{sc}t}{G_p}) \ P_1$$

where it should be noted that $q_{sc}t/G_p$ is the fractional gas recovered. If we assume a value for P_{avg} and substitute this value in equation 115, then we can obtain a production possibility set assuming a constant flow rate q_{sc}. We plot this production set in Figure 19 with pressure and cumulative gas recovery noting that as P_{avg} declines cumulative recovery increases until some arbitrary pressure p_a, the abandonment pressure, below which it is impossible to obtain any further primary recovery.

Finally before giving an example, we would like to arrive at the production function for the case of gas recovery. We will plot well radius which has a high degree of substitution in terms of equipment, pipe radius, as a function of capital, against the rate of production which is a function of time. The relationship between capital and time is that the larger the well radius the more capital invested, and thus the greater the production rate and vice versa. Economics of scale also make it economical to install a greater pipe radius than demand would call for if an increase in demand for gas is anticipated. This isoproduct curve is given in Figure 20.

Finally, flow production rate with reserves considered, can be given by:

135

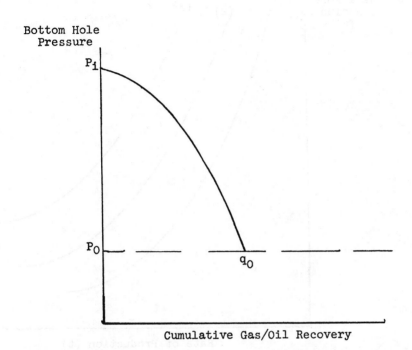

Figure 19   A Production Decline Curve for Gas/Oil
Reservoir

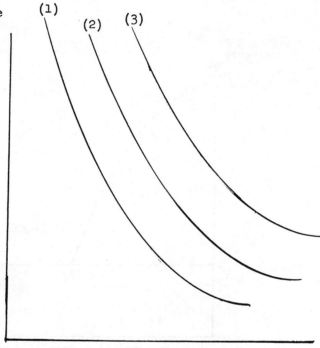

Well Bore
Radius
(K)

Rate of Production (t)

Figure 20   A Set of Isoproduct Curves for
Gas/Oil Reservoir

Note: The isoproduct curves are drawn for
various well bore radius, and thus
for various amounts of capital
employed. Obviously, as we move
outward more capital is being
employed to obtain higher rates of
production.

$$116 \qquad q_{sc} = \frac{703kn(1 - \frac{q_{sc}t}{G_p})^2 \ (P_1{}^2 - P_w{}^2)}{\mathcal{M} T \ \ln(.472) \ r_e/r_w}$$

We will be concerned later on in finding the optimum flow rate given that the constraining factors are satisfied. Now we would like to utilize some of the tools that have been presented in this analysis to consider a hypothetical gas and oil reservoir, and to see if we cannot apply these techniques to these hypothetical well problems.

Practical Application Gas

Suppose that we wanted to calculate the gas recovery from a gas reservoir with following specifications:

A = 640 acres;

h = 22 ft. (thickness);

$\emptyset$ = .15;

P_1 = 3000 psia;

T = 140° F.;

$\mathcal{M}$ = .015 centipoise;

Z = 1.00 (ideal gas);

r_w = .333 ft.;

First let us assume that the one well is drilled in the middle of the 640 acres. If k = 5 millidarcys and q_{sc} = 2.0 MMCFD, then the external radius is given by:

117
$$r_e = \sqrt{\frac{640 \times 43{,}560}{3.1416}}$$

or:

$$r_e = 2{,}980 \text{ ft.}$$

The initial gas in place at initial pressure P_i for a reservoir with an ideal gas is:

118
$$G_p = 3.1416\ (2{,}980)^2\ 22(.15)\ (\frac{3000}{14.7})\ \frac{520}{600}$$

$$G_p = 16.27 \times 10^9\ \text{SCF}$$

For our ideal gas, the gas remaining at any time will be directly proportional to the average pressure:

119
$$P_{avg} = (\frac{1\text{-}2 \times 106\ t}{16.27 \times 109})\ 3000$$

$$P_{avg} = 3000 - .369\ t.$$

Then we can substitute this into the following equation:

120
$$\frac{703\ kh}{\mu\ TZ\ \ln\ (.472\ r_e/r_w)}$$

$$= \frac{703\ (.005)\ 22}{.015\ \ln\ (.472\ \frac{2980}{.333})\ 600}$$

$$= 1.029$$

and substituting into:

121
$$q_{sc} = \frac{703\ kh\ P_{avg}{}^2 - P_w{}^2}{\mu TZ\ \ln\ (.472\ r_e/r_w)}$$

we obtain:

122       $q_{sc} \; \mu TZ \; \ln .472 \; (r_e/r_w) \quad = 703 \; kh$

$$(3000 - .369t)^2 - P_w^2 \quad 2 \times 10^6$$

$$= 1.029 \; (3000 - .369t)^2 - P_w^2.$$

If we allow P_w, bottom-hold pressure, to vary, we can find
time, t, in days. So we can find the recovery at the end
of any time period, for example, if P_w = 600 psia, then
substituting in the above equation we find that t ≃ 4,016
days or 11 years. The recovery factor at the end of 11
years would be:

123   R.F. ≃ $\dfrac{q_{sc}t}{G}$

$$= \frac{2 \times 10^6 \; (4016)}{16.27 \times 10^9}$$

$$≃ \frac{8.02 \times 10^9}{16.27 \times 10^9}$$

≃ .493 or 49.3 at the end of 11 years.

And if we plotted the above information, we could obtain
our production possibility set which is given in Figure 21
Suppose that we changed the production rate from 2 MMSCF/day
to 1 MMSCF/day and leave all other conditions in the problem
the same. We want to see what effects this change will have
on the problem. Then for P_w ≃ 600 psia, we would find the
time factor to be:

140

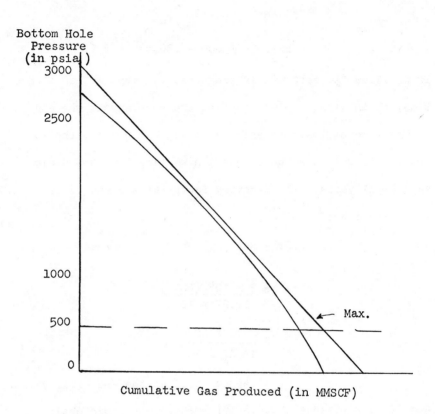

Cumulative Gas Produced (in MMSCF)

Figure 21   A Production Decline Curve for a Gas
Reservoir Assuming K = 5 MD, Q
= 2MMSCF/day and 1 well per 640 Acres

124 $\qquad$ $1 \times 106 = 1.029 \quad (3000 - .1845t)^2 - 600^2$

$t = 10,010$ days or 27.4 years

The recovery at the end of 27.4 years would be:

125 $\qquad$ R.F. $= \dfrac{q_{sc}t}{G}$

$$= \frac{1 \times 10^6 \cdot 10,010}{16.27 \times 10^9}$$

$$= \frac{10 \times 10^9}{16.27 \times 10^9}$$

$$= 61.5\% \text{ at the end of 27.4 years.}$$

The production possibility set for this new problem is given by the line labeled Case II in Figure 22. The previous possibility set is labeled as Case I

In Figure 22 the maximum gas recoverable would be given by the line labeled MAX, where the gas recovery is 16×10^9SCF of gas, the Case II line being the gas produced if $k = 5$ MD and $Q = 1$ MMSCF/day or a well-spacing of 640 acres, and the third line being the production set for the initial conditions. We could, in Figure 22, indicate other changes which might affect recoverability, changes in permeability, porosity, viscosity, well radius, etc., to indicate the production possibilities available to the firm developing the well under these changing conditions.

As another possibility, we could vary the number

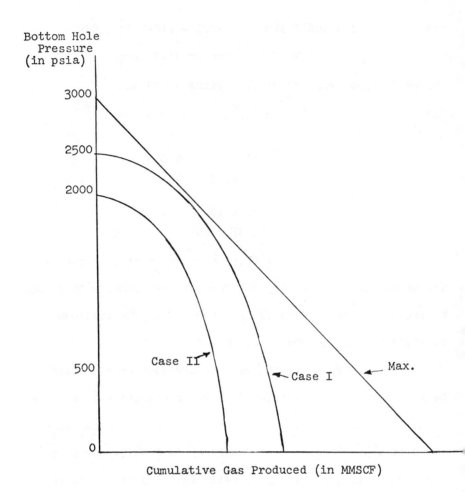

Figure 22 A Production Decline Curve for a
Gas Reservoir Assuming K = 5 MD,
Q = 2MMSCF/day and 1 well per
640 Acres

Note: Case I   Case II

           K = 5 MD   K = 5 MD
           Q = 2 MMSCF/day   Q = 1 MMSCF/day
           1 well per 640   1 well per 640
             acres   acres

and the spacing of the wells, for example, let us consider a well-spacing of 160 acres between wells and four wells on the 640 acres. Then if $k = 5MD$, $q_{sc} = 2.0$ MM CFD or 8.0 MMCFD for the four wells, the external boundary radius would be:

126
$$r_e = \sqrt{\frac{160 \times 43{,}560}{3.1416}}$$

$$r_e = 1{,}490 \text{ ft.}$$

and

127
$$= \frac{703 \text{ kh}}{\mu T \ln (472 \ r_e/r_w)}$$

$$= \frac{703 \times .005 \ (22)}{.015 \ (600) \ln (.472 \ \frac{1490}{.333})}$$

$$= 1.122.$$

Then

128
$$\frac{q_{sc}t}{G} = \frac{2.00 \times 10^6 t}{\frac{16.27}{4} \times 10^9}$$

$$= 1.476t$$

and

129
$$2 \times 10^6 = 1.122 \quad (3000 - 1.476t)^2 - P_w^2$$

If we again allow $P_w = 600$, then:

130
$$2 \times 10^6 = 1.122 \quad (3000 - 1.476t)^2 - 600^2$$

$$t = 1{,}040 \text{ days or } 2.85 \text{ years.}$$

At the end of 2.85 years the recovery factor would be:

$$= \frac{q_{sc}t}{G}$$

131 $$= \frac{2 \times 10^6 \times 1040}{(\frac{16.27}{4}) \, 10^9}$$

$= 51.1$ recovery at the end of 2.85 years.

Thus we see the effect of closer spacing on development time of the well. We will consider this problem in more detail when we consider the capital costs incurred in producing those four wells to determine the economic efficiency of developing the well.

The production possibility set for these new sets of conditions we have proposed is given in Figure 23. Note that the initial pressure is higher than Case I and the cumulative recovery is also higher than in the first case analyzed, but lower than in the second case when we changed Q to 1.

There are many other combinations or sets of combinations which we could utilize. In general, we can summarize what we have done here by saying that the greater the well-bore radius, the faster production is possible; the more permeable the rock, the faster the flow rate and thus faster production is possible; the greater the porosity the greater the allowable flow rate and thus the faster

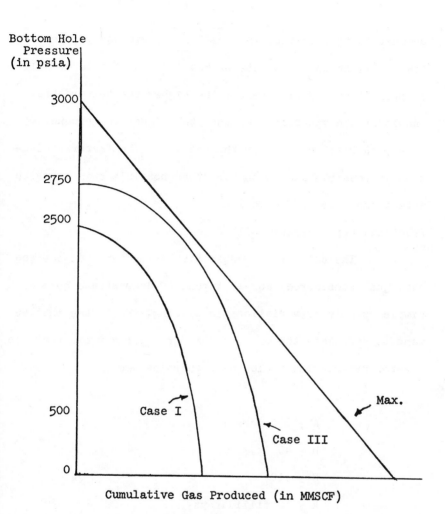

Figure 23   A Production Decline Curve for a
            Gas Reservoir Assuming K = 5 MD,
            Q = 2MMSCF/day and 4 wells

Note: Case I is as before.
            Case III

            K = 5 MD
            Q = 2 MMSCF/well or 8 MMSCF Total
            4 wells equally spaced at 160 acres
              apart on 640 acre plots

production is possible; the greater the viscosity, the
lower will be the flow rate of gas and thus the lower will
be production and vice versa; the higher the temperature,
the lower the viscosity and thus the higher the productive
rate possible; the greater the pressure, the faster production is possible and so on for other possible changes which
affect the rate of production.

Practical Application: Oil

The case of oil estimation is easier than the gas
case just considered, and we intend to demonstrate here a
simple example of variations in spacing. We assume that we
want to calculate the oil recovery from a reservoir that has
a water drive. The following properties are

$$A = 640 \text{ acres};$$
$$h = 40 \text{ ft.};$$
$$\emptyset = .10;$$
$$k = 10 \text{ millidarcys};$$
$$S_o = 307°;$$
$$t = -140° \text{ F.};$$
$$P_i = 2,000 \text{ psig};$$
$$B_{oi} = 1.388;$$
$$R_s = 730;$$

$$B_g = .0015;$$

$$\mu_o/\mu_f = 35;$$

From the material-balance equation, it is possible to find the initial oil in place for the reservoir.

132
$$N_1 = \frac{7758 \; Ah\phi \; (1-S_o)}{B_{oi}}$$

$$N_1 = \frac{7758 \quad 640(40).10(1-.30)}{1.388}$$

$$N_1 \overset{\backsim}{=} 13,900,000 \; STB$$

originally in place on the reservoir.

Assuming we know P_W and we have the initial pressure P_1 and the external radius, and assuming the well radius to be $r_W = 1$ ft., then we can obtain the rate of production for the reservoir:

133
$$q_{sc} = \frac{7.08 \; kh \; (P_1 - P_W)}{B_o \; \mu[\ln r_e/r_W - \frac{1}{2}]}$$

$$q_{sc} = \frac{7.08 \quad 10(40)(2000-1000)}{35(1.388 \; [\ln 2980 - \frac{1}{2}]}$$
$$1$$

$$q_{sc} = 235 \; STB/day$$

at the initial pressure of $P_1 = 2000$ psig.

Suppose production begins with this one well. If we assume a pressure drop, we want then to find the average pressure when P_e is say 1500 psig. Then the P_{avg} can be given by:

134
$$P_{avg} = P_w - \frac{\mu q_{sc} B_o}{7.08 \ kh} \left[\ln \frac{r_e}{r_w} - \tfrac{1}{2} \right]$$

$$P_{avg} = 1896 \ psig$$

For $P_w = 1898$ we found the q_{sc} at that pressure to be:

135
$$q_{sc} = \frac{7.08 \ kh \ (P_{fw} - P_w)}{\mu B_o \left[\ln r_e/r_w - \tfrac{1}{2} \right]}$$

$$q_{sc} \cong 226 \ STB/day$$

Using the material-balance equation, we find that at this pressure there has been a recovery of

136
$$N_1 = \frac{B_{o1} N_p}{B_{o1} - B_o}$$

$$N_p \cong 50,000 \ STB$$

which in terms of percentage of initial oil in place can be found by:

137
$$R.F. = \frac{N_p}{N_1}$$

R.F. $\cong$ .004% of the initial oil in place. To find the time required to produce the N_p we use:

$$t = \frac{N_p}{q_{sc}}$$

$$t = \frac{50,000}{226}$$

$$t \cong 221 \ days$$

Let us keep the same properties of the reservoir, only now allow four wells to be placed on the initial reservoir tract. Again using the same P_w the external radius will now change, specifically $r_e = 1490$ ft. assuming no interference. Thus the rate of production for the reservoir will now be:

139
$$q_{sc} = \frac{7.08 \quad 10(40)(2000-100)}{35(1.388) \left[\dfrac{\ln 1490 - \frac{1}{2}}{1} \right]}$$

$$q_{sc} \cong 655 \text{ STB/day}$$

or approximately 164 STB/day per well at the initial pressure of $P_i = 2000$ psig.

Assume production continues and we went to find the average pressure, P_{avg}, when $P_e = 1500$ psig. Then as we have seen previously $P_{avg} = 1896$ psig. At this pressure however, q_{sc} will now be 620 STB/day or 155 STB/day per well. The time required to produce this 50,000 STB of oil however, is:

140
$$t = \frac{5000}{620}$$

$$t \cong 80.6 \text{ days}$$

as compared to 221 days for the prior case.

If we can estimate the percent of feasible recovery, which would not be difficult to estimate from the pressure data, we could obtain an estimate of the percentage of reservoirs recoverable, and then determine the best method of spacing.

Let us assume for purposes of illustration that the abandonment pressure is $P_a = 200$ psig. At this new pressure conditions in the reservoir have changed. Assume now that

141
$$B_{oa} = 1.100;$$
$$\mu_g = 150;$$
$$k = 15;$$

Then the rate of production q_{sc} at abandonment pressure for a single well can be given by:

142
$$q_{sc} = \frac{7.08 \ kh \ (P_e - P_w)}{B_o \ \mu \left[\ln r_e/r_w - \frac{1}{2} \right]}$$

$$q_{sc} = \frac{7.08 \quad 15(40)(200-100)}{150(1.10) \ \left[\frac{\ln 2980 - \frac{1}{2}}{1} \right]}$$

$$q_{sc} \cong 5.4 \ \text{STB/day}$$

at abandonment pressure. At this pressure, the amount produced or cumulative production is:

143
$$N_1 = \frac{B_{oi}N_p}{B_{oi} - B_o}$$

$$N_p \cong 2,164,000 \ \text{STB}$$

The time required to produce this output from a single well
is roughly estimated by finding the average production per
day over the life of the well and dividing by the cumulative
production. Assume the q_{sc} avg. $= 75$ then:

144
$$t = \frac{2,164,000}{75}$$

$$t \approx 30,186 \text{ days or } 85.4 \text{ years.}$$

for this well to produce to abandonment pressure.

If we assume four wells as previously, then the
new q_{sc} we obtain at abandonment can be given by:

145
$$q_{sc} = \frac{7.08 \left[15(40)(200-100) \right]}{150(1.10) \left[\ln 1490 - \frac{1}{2} \right]}$$

$$q_{sc} = 17 \text{ STB/day or } 4.25 \text{ STB}$$

per day per well. The recovery will be the same as we found
above or about 16% of the initial oil in place. Only now
that four wells are in operation the average rate of production for the reservoir will be appreciably higher and
thus the accumulative time to produce this primary recovery
decreased considerably. If we assume the average production
rate to be 250 STB/day for the reservoir, then:

146
$$t = \frac{2,164,000}{250}$$

$$t \approx 8,656 \text{ days or } 20.9 \text{ years.}$$

Thus we can see the effect well spacing can have on the time recovery pattern of a given reservoir, making the project more feasible in terms of life of the project.

If we wanted, we could extend this analysis to consider changes in k, μ , B , and the other variable parameters of the system which would change over time and give us changes in the production rate, which in turn would lead to new time patterns. However, this is only a simple hypothetical example of the effects of well spacing on a reservoir, and we need not consider such variations.

Summary

The purpose of this chapter has been to expose the two relevant problems that need to be considered when considering the physical aspects of reservoirs recovery, namely the calculation of reserves, and the flow of fluid in a porous media. We will use these principles and equations derived further when considering physical ultimate recovery ane economic ultimate recovery. However, presently we would like to consider the economic aspects of the recovery process, the determination of the interdependent problems of rate of production, ultimate recovery, and well spacing from a diagramatic economic standpoint.

CHAPTER 5

ECONOMICS OF WELL SPACING: A DIAGRAMMATIC

PRESENTATION OF THE RELEVANT

ECONOMIC THEORY

Introduction

It was not until the last of the 1940's that the
problem of well spacing was perceived as predominately an
economic problem, and it was at least another decade before
a full treatment of the subject was undertaken, by an engineer, and the economic impact of the problem realized.

Economists per se, and the economic profession
have never, until recently, been interested in the problem
of well spacing. However, as a result of growing concern
with the problems of natural resources during and particularly after the Second World War, economists unawarelly
backed themselves into a position where the topic needed
discussion. One of the first post-war pieces of work was

by Eugene Rostow.[1] It was the first piece of work done by
an economist delineating the petroleum conservation problem
and policy which might be implemented to conserve the
natural resources of petroleum crude. It was not until 1955
and the work of H. Anthony Scott[1] that a full definition of
the conservation problem in economics was undertaken and the
problem properly placed within the theory of capital.

Basically, in his work, Scott is concerned with
all natural resources rather than just petroleum, but his
general definition of the conservation problem and its place
in economic theory transcends the specific applications that
were undertaken in the book. Natural resources are classified by Scott into two groups: those that are replenishable
such as fish, timber, and other regenerative species, and
those that are nonreplenishable such as oil, gas, and other
mineral products. The latter resource represents then a
fixed stock of assets to the economy at any point in time,
and thus efficiency in the utilization of these resources
becomes important since over time the existing stock can

---

[1]Eugene Rostow, A National Policy for the Oil
Industry, (New Haven, Yale University Press, 1948).

[2]H. Anthony Scott, Natural Resources, The
Economics of Conservation (Toronto, University of Toronto
Press, 1955).

only diminish with use.

Many examples of the misuse of both replenishable and nonreplenishable resources exist, witness for the former the depletion of the wild bison, or the fishing stock in the Great Lakes. However, the regenerative resource always represents the hope or at least has the capacity for regeneration if man cooperates with nature. The nonreplenishable resource, on the other hand, represents a stock of resources, an inventory which can only decrease, and when a faulty time distribution of resources occurs, these resources are lost to society for good.

Here obviously the concern is with the misuse of these nonreplenishable stock of resources. In this regard, the question of conservation has been very much discussed and debated in economics in the past two and one-half decades. Two dominant schools of thought have emerged in regard to this question; those advocating the non-use or the substitution of replenishable resources for nonreplenishable resources where possible; and secondly, those advocating the efficient management and recovery of nonreplenishable resources.

The view taken here is with the efficient management and recovery of natural resources, that is, the application of modern scientific management techniques to

the efficient recovery and production of gas and oil

resources. Thus, areas of concern in this treatment will

be with determination of the rate of production, the econo-

mic ultimate recovery, and well spacing patterns for a

reservoir to achieve this efficiency. As will be seen later,

all these problem areas are interdependent.

Economics of Conservation and Maximum Efficient Rate
of Production

That the domestic petroleum industry has been

characterized by wasteful production will not come as a

surprise to anyone acquainted with the present prorationing

system in the various producing states, or to anyone

acquainted with late nineteenth and early twentieth century

American oil production history.[3] Unanimously, investi-

gators in the area have agreed that some controls are

necessary at the production level. As to what form these

controls should take in terms of public policy measures is

---

[3]C. W. Zimmerman, Conservation in the Production
of Petroleum, (New Haven, Yale University Press, 1957),
pp. 140-184 has a good historical development of the
prorationing system in the producing states.

still at this late date very much in question.[4] The
traditional arguments have been in terms of some type of
mandatory or voluntary production control which would limit
supply to demand at the going price level. Others have
advocated over time, the voluntary or, if necessary, the
mandatory unitization of reservoirs, that is, the operation
and development of a given reservoir as a single unit of
production. Which system should be recommended as a policy
measure depends on the definition of efficiency. However,
it is still realized that huge inefficiencies continue to
exist in the present production system, and that modification of the existing public policy both federal and state,
with regard to the prorationing system is still a necessity.

Since we are dealing with an industry and firms
in that industry, it would appear that the traditional
economics of the firm should be sufficient for our considerations of industry behavior. Traditionally, the basic
demand, supply analysis is the methodology used for
determining what the optimum decision of an individual firm

---

[4]As recent as 1960, a call to modify the present
prorationing system in all producing states has been advocated, (1) by engineers, (2) by economists, (3) by large
producers. The only nonsupportive group has been the small
marginal producers and their interests have dominated until
this date. See Oil and Gas Journal, January 18th, 25th, 1960,
pp. 42 and 55.

or group of firms will be with regard to equilibrium quantities produced and equilibrium prices charged. However, this traditional Marshallian analysis has many drawbacks when considering a natural resource, namely the fact that it is static analysis. However, it is not fair to contend that time is circumvented by contrasting long-run and short-run equilibria situations. While these devices are useful in determining output and prices under the stated assumptions, they explicitly avoid the determination of the time distribution of resources, or output, the most important factor in determining the equilibrium level of output for natural resources.

Another important variant with the traditional theory is that the demand-supply analysis is solely concerned with the determination of output. However, with natural resources, the important question is determining the rate of utilization of a fixed factor of production, since the stock of resources is nonrenewable. Thus, in order to overcome these deficiencies imposed by traditional theory, it is necessary to study production, ultimate economic recovery, and well spacing in the context of capital-investment theory.

There are many possible ways to define production efficiency, and it is important here to review some of the

more common definitions. In the conservationalist's sense,
efficiency can be defined in terms of the efficient time
use of resources. The problem for the firm then is to
optimize the existing stock of resources over time. Since
production is a function of time, or the time use, and since
this time use of the stock of resources is being optimized,
then the rate of production which will optimize the time
use of resources is the optimal efficient rate of production.
This in a social sense is one possible definition of maximum
efficient rate of production. However, in the gas-oil
context it is not often used.

It is also possible to view MER from the point of
view of the firm. Here, the MER may simply be defined as
that rate of production at which the discounted net revenue
-- gross revenue minus expenses for the current factors of
production -- earned from the last unit produced, is equal
to the net discounted revenue which the unit could have
earned if its production were postponed into the future.
The discounting of these returns to the present must be,
of course, undertaken at the marginal rate of time preference for the individual firm. If there exists a perfect
market for funds, then the discounting can be undertaken at
the "equilibrium rate of interest." This alternative return

which is expected to be available in the future as a result
of present nonuse is often referred to as the opportunity
cost, or the user cost of production today. Thus, a pro-
ducer making decisions regarding output for a natural
resource should not be thought of as equating marginal cost
and marginal revenue in the traditionalists sense, but
rather as equating his marginal net revenue and his marginal
users costs which are both a function of time and the discounting factor. When MNR = MUC then the firm can be said
to be maximizing the present value of its product and thus
producing at MER.

The concept MER has in the past had two connota-
tions, one physical and the other economic. Of course, the
MER for any given reservoir depends somewhat on the peculiar
engineering and physical characteristics of the reservoir.
We have discussed this in some detail in Chapter 2 in its
historical context, and in Chapter 3 in its engineering
sense. However, the MER is not a physical maximum, but has
economic meaning, and this economic meaning defines the
spacing problem in its proper context. The following quota-
tion defines the MER in an economic context:

Operation of a reservoir at an infinitesimally low rate of oil production would not in itself assure efficient recovery unless other necessary conditions were met. However if all other conditions are fulfilled, the ultimate oil recovery from most pools is directly dependent on the rate of production. The nature of this dependence is such that for each reservoir there is for the chosen dominant mechanism a maximum rate of production that will permit reasonable fulfillment of the basic requirements for efficient recovery. Rates lower than such maximum may permit still higher ultimate oil recovery, but once the rate is sufficiently low to permit the basic requirements to be met, the incremental ultimate recovery obtainable through further reduction of the rate of production may be insufficient to warrant the additional deferment of a return and additional operating expenses that would result from a prolongation of the operation. A rate of production so low as to yield no return would obviously be uneconomic and of no ultimate benefit. However, increase of the rate of production beyond the maximum commensurate with efficient recovery will usually lead to rapidly increasing loss of ultimate recovery. From these considerations there has developed the concept of the maximum efficient rate of production, often referred to as the MER. For each particular reservoir it is the rate which if exceeded would lead to avoidable underground waste through loss of ultimate oil recovery.

This definition we shall use for MER from now on.

The Economics of Production

In the problem of production the first important problem that has to be answered is the type of drive to use.

_____

[5]S. E. Buckley (ed.), Petroleum Conservation publication of The American Institute of Mining and Metallurgical Engineers, (New York, 1951), pp. 151-152.

The physical characteristics of the reservoir will determine
the type of drive for a given system. There are basically
three possible types of drive, dissolved gas, gas cap, and
water drive, plus possible combinations of these three.
We would like at this point to discuss these drives and
their characteristics and efficiency. What is said here can
be found also in the 1943 Standard study.[6]

In the case of dissolved gas, the only displacing
agent is the gas released in solution, since there is no
other source of gas or water. The process is inherently
inefficient because the dissolved gas is released everywhere
throughout the reservoir and thus cannot be prevented from
escaping along with the oil through the producing well(s).

The mechanism of gas cap drive, on the other hand,
requires continuous pressure maintenance throughout the
recovery process and a distinct segregation between an
enlarging gas-invaded zone containing reduced oil saturation,
and a shrinking oil zone containing high oil saturation.[7]
Since comparatively little oil is usually produced after the

---

[6]Joint Progress Report on Reservoir Efficiency and
Well Spacing, Committee on Reservoir Development and Operation of the Standard Oil Company (New Jersey) and Affiliated
Companies of the Humble Oil and Refining Company (Dallas,
1943).

[7]Ibid., p. 25

invading free gas has completely penetrated the oil zone,
the ultimate oil recovery is very nearly proportional to the
degree of oil desaturation of the reservoir at the time the
advancing gas front has reached the lower-most wells.

The recovery efficiency by this drive mechanism
is very sensitive to the rate of oil production. Excessively
high rates of production would cause rapid encroachment of
the free gas throughout the oil zone with a relatively low
displacement efficiency. After, it would be difficult to
maintain segregation of free gas, and the entire free gas
content of the reservoir would be dissipated and thus pressure reduced, and recovery more difficult since pressure is
the dominating recovery mechanism of the gas-cap drive.

An efficient rate of production under gas-cap
drive must be a rate such that gravity will dominate the oil
flow to a sufficient extent to maintain continuously an
advancing gas front behind which the oil saturation throughout the formation will be reduced to a satisfactorily low
value, in regions of low permeability as well as in highly
permeable regions. This results in producing at such a rate
that oil migrates into the lower portion of the reservoir
and maintains the oil saturation therein by gravity drainage
instead of being compelled to migrate by expanding gas

forcing its way into the oil zone in response to a pressure differential between high pressure in the gas-cap and low pressure in the oil zone. The actual pressure in all parts of the oil zone must remain higher than the pressure in the gas-cap, and the free gas must be relegated to the role of merely expanding to fill the space vacated by the downwardly migrating oil. The chief function of the gas then is to maintain the pressure level at which the gravity drainage proceeds.

When water drive is the dominant force in the reservoir, then other considerations must be accounted for in determining the maximum efficient rate of production. First, the net water influx into a reservoir must volumetrically be substantially equal to the net oil withdrawn. Water can move into a reservoir only as a result of a pressure differential between the aquifer zone and the oil zone. For any given reservoir configuration and formation permeability, the rate at which water can invade the reservoir at any point in time is directly proportional to the pressure differential between the aquifer and the reservoir. The faster the oil production, the higher must be the pressure differential between the aquifer and the reservoir for the water influx to keep pace with the oil withdrawal.

The determination of the maximum efficient rate of production for a water drive reservoir requires two necessary circumstances: first, the determination of the proper level or proper range of reservoir pressure to be maintained throughout the production process; and secondly, the calculation of the rate of influx of water at the predetermined level of reservoir pressure maintenance.

The proper pressure is usually taken to be that one which will not permit dissolved gas to be released in sufficient quantity to build up within the oil zone a free gas saturation large enough to allow flow of the liberated gas. Because of the expansible nature of reservoir fluids and of the water in the aquifer, adjustments in reservoir pressure do not take place instantaneously with adjustments of withdrawal rates; as a practical matter, it is usually necessary to determine the level to which the reservoir pressure will be allowed to decline eventually over the life of the production. Thus, determination of the efficient rate of production of a water-driven reservoir is a relationship between the reservoir pressure and the rate of water influx.

As in the case of the gas-cap drive, it is necessary to provide reasonably uniform encroachment of the water-oil interface, and uniform displacement of the oil

behind that interface in the regions permeated by the water.

Control of the uniformity of the advancing water flow is

dependent on the balance between the component of gravity in

the direction of the flow and the pressure gradients induced

by the flow. Since the density contrast between oil and

water is less than the density contrast between oil and gas,

control of the uniformity of the advancing water front is

more difficult to control, and it requires that the pressure

gradients induced by the flow be even less than in the case

of the gas-cap. Thus, a third necessary requirement for the

MER to be operable is that the rate of production be such as

to allow gravity to regulate the level of the advancing

water front. Partially offsetting this requirement is the

fact that water is more thorough flushing agent than is gas,

and thus much higher degrees of efficiency in drainage are

attainable with water.

Because of variations in permeability of the

reservoir rock, there is a natural tendency for the flow of

both water and oil to take place primarily in the more

permeable channels of the rock. However, most reservoir

rocks are preferentially wetted by water in the presence of

oil; thus, the capillary forces set up by rock texture and

interfacial tensions are such that it causes water selectively to enter the finer interstices and regions of lower
permeability and to eject the oil from these regions into
the more permeable sections. These capillary forces are
very small and do not cause rapid movement except over very
short distances. If an area of a reservoir is to be flushed
by water, the rate of advance of the water must be slow
enough to permit the water selectively to eject the oil from
the less permeable into the more permeable channels. Thus,
the important fact in water drive is that the rate of water
advance be limited so that the water can uniformily and to
a high degree displace the oil from all portions of the
reservoir rock.

Thus, the control of the rate of production in a
water drive reservoir has three important aspects: first,
the rate of production must be controlled to the degree
that the oil may be volumetrically replaced by the water at
the proper pressure maintenance; secondly, the rate of production must be controlled to such a degree that the force
of gravity may keep fairly uniform the advancing oil-water
interface; thirdly, the rate of production must be controlled to the degree that the advancing of the water with
the advance of the interface selectively eject the oil from

the less permeable channels into the more permeable channels
so that the reservoir may be flushed more uniformally and
complete. Given these conditions are approximated, then to
some degree efficiency is attained in a water-drive reservoir.

The Economics of Well Spacing

The concern with these different types of drives
is to determine the relative efficiency of the drive in
question. Also, within the limits of the drive, it is then
possible to determine the maximum efficient rate of production for a reservoir, and more directly and importantly,
the maximum cumulative production for a reservoir given the
production rate and time depletion factor.

If the assumption of a constant rate of production
(MER) is made over the life of a reservoir, and if recovery
by secondary means is ignored, then it is possible to
determine over time the maximum cumulative production from
a reservoir given the well spacing. In Figure 24 , we have
depicted the case of maximum physical cumulative recovery.
We have plotted time on one axis, and total physical
recovery on the other. We see that as we allow for a
greater time span of recovery, the greater is the ultimate
recovery until theoretically it attains the maximum recovery

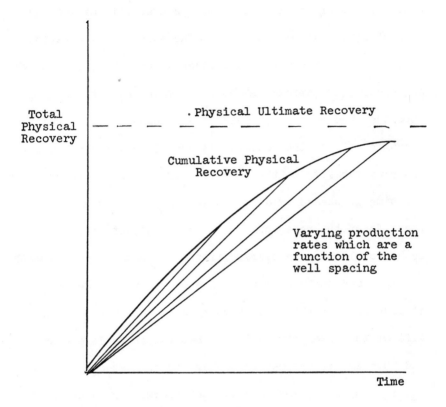

Figure 24   Physical Ultimate Recovery

possible for the reservoir by natural drive. While physically it may be feasible to operate a well and a reservoir to the point of physical ultimate recovery, it may not be economically feasible to do so. The economic life period of a reservoir needs to be considered, and thus, the important concept of maximum economic ultimate recovery is more important and relevant. This is so because, first, the economic life of the reserves is considered, and secondly, the profitability of the operation of the reservoir over its lifetime is heeded by discounting costs and revenues over the life of the reservoir. Figure 25 depicts the total economic ultimate recovery. First, it assumes a given well spacing, and unchanging technique of recovery. Secondly, it assumes that the production rate is constant over the life of the reservoir until the natural drive is exhausted; and thirdly, it assumes that there is no secondary recovery. P_1, P_2, P_3, etc., indicate various production rates for the reservoir. The relation between the production rates is that the lower the production rate, the larger is the fraction of total reserves recovered, and the longer the time period of recovery, and thus the longer the economic life of the reservoir and the investment.

Having considered the problem of production rates,

171

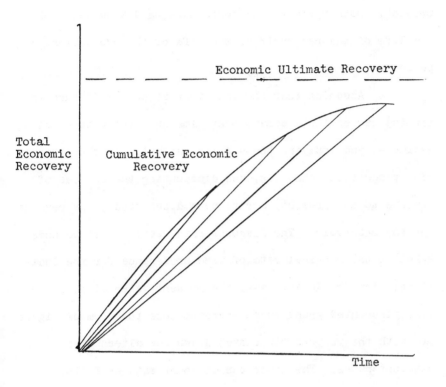

Figure 25 Economic Ultimate Recovery

it is next essential to consider the revenues received from
production, and the costs incurred in the production of the
product. Both revenues and costs must be discounted over
the life of the reservoir or the life of the production
period.

Assuming that the reservoir is unitized, then the
firm(s) can estimate demand over time and arrive at a price
estimate, and applying these price estimates to the alternative production schedules, and discounting the sources of
revenue to the present, arrive at a discounted gross revenue
for the reservoir. The discount rate used: "must be undertaken at the marginal rate of time preference for the individual firm."[8] In any case, the present value of alternative discounted gross income streams take the form of Figure
26 with the various DGR curves drawn for alternative
discount rates. The lower the discount rate used, the
higher the DGR curve; the higher the discount rate, the
lower the DGR curve. Thus, we have discounted gross
revenues over time. It is also possible to plot discounted
gross revenue with well spacing, the alternative DGR curves
again being drawn for alternative discount rates used. The

---

[8]See Chapter 2 p. 46 of the present thesis.

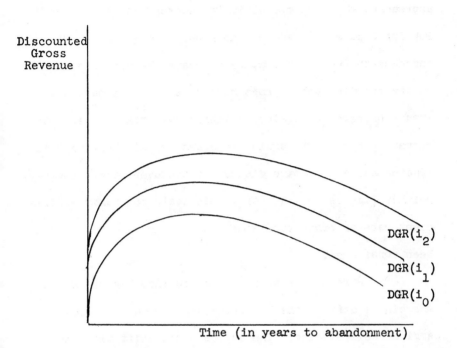

Figure 26   Discounted Gross Revenue Curves

Note: i is interest rate and $i_0 > i_1 > i_2$.

curves indicate that as more wells are drilled less and less is added to discounted gross revenues until it finally approaches the economic ultimate recovery as in Figure 27 Putting Figures 26 and 27 together we see that as we increase the well spacing, we decrease the time required to obtain the discounted gross revenue and we decrease the amount of revenue obtained slightly over time. These DGR curves in Figure 28 again are drawn for alternative well spacing and alternative discount rates over time. However, neither Figures 26 , 27 or 28 indicate the maximum efficient rate of production since only discounted revenues have been considered.

However, it should be noted that Figure 28 and wherein we defined the EUR curve and the DGR are drawn for a given well spacing. The curves would shift with changes in the spacing.

As we have seen earlier, the physical relationship for each reservoir is different. However, it is possible to diagram its general form in Figure 29 . In this diagram, we have reservoir production rates on the ordinate axis and well spacing on the abscissa. Each U curve represents a different physical ultimate recovery, and indicates the combinations of production rates and number of wells consistent with that

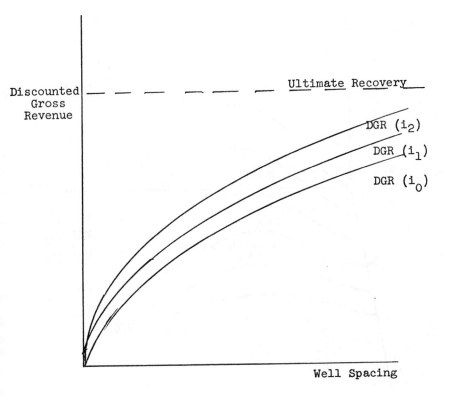

Figure 27   Discounted Gross Revenues and
Well Spacing

Note: i is interest rate and $i_0 > i_1 > i_2$.

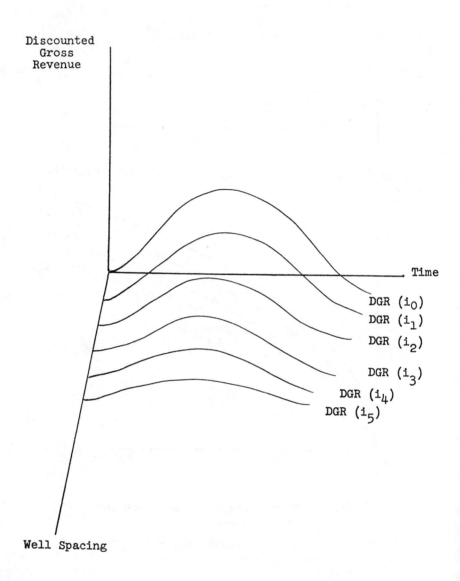

Figure 28   Discounted Gross Revenues, Well
Spacing, and Time

177

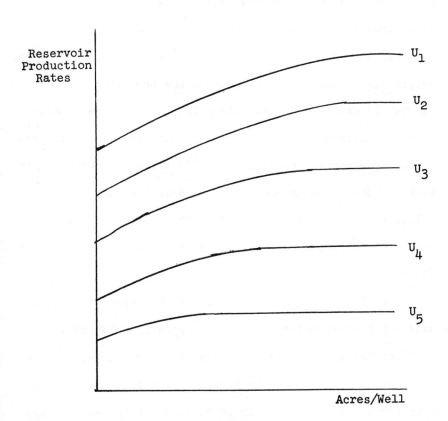

Figure 29   Reservoir Production Rates and
Well Spacing

given ultimate recovery. Increasing physical ultimate recovery is indicated by a downward shift of U_1 to U_2 to U_3 etc. The various production rates as given in Figure 25 would then correspond to the reservoir production rates arrived at in Figure 29 . The curves do not originate at the origin since once a well is drilled and producing, it is assumed to have a positive production rate. There is also a physical maximum reservoir rate of production, a point where the U curve becomes horizontal. This indicates that you can increase the ultimate recovery by increasing the number of wells, but only to a point, and at that point the maximum rate of production for that number of wells for the reservoir is attained.

For each number of wells on the reservoir, there exists a corresponding U curve, and as well a DGR curve. Plotting the alternative DGR curves for various well spacing and over time yields us the present value of gross revenues over time. As the reservoir production rate rises with any given well spacing, the present value of gross revenue rises at first because of the shift of income into the present, but eventually falls because of the adverse effects of higher production rates on ultimate recovery. In Figure 30 each PUGR curve is drawn for a given ultimate recovery curve,

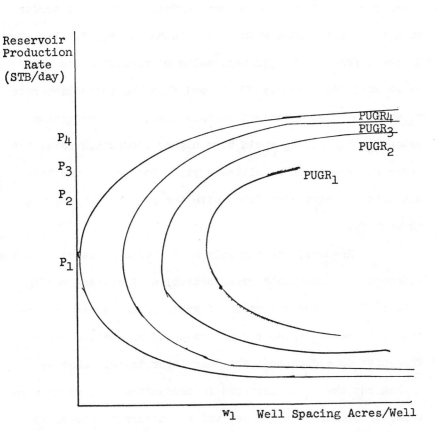

Figure 30   Present Value of Gross Revenue Curves

and for a given discount rate. Assume that we have well spacing given by w_1. Then there exists an array of production rates associated with this spacing P_1, P_2, P_3, P_4, etc. However, for w_1, the greatest value of present gross revenues would be given by $PUGR_1$, and thus the production rate P_1 would be most efficient in this case. The production rates P_2, P_3, and P_4 would have higher production rates but lower present value of ultimate yield because of the less ultimate recovery associated with the production rates P_2 through P_4.

However, looking only at the present value of gross revenues will not yield us an optimal well spacing configuration. To the present value of gross revenues, it is necessary to take out the present value of production costs. The costs considerations are seen later in this chapter. Taking out the costs incurred in production will yield a net value of product, or the net value contours of Figure 31. Here we see that the benefits of adding more wells diminishes until the contours of the net present value form a maximum at M. Given the present value of gross revenues less production costs, the point M will give us the optimal reservoir production rate (MER) and the optimum number of wells for the reservoir. In Figure 31, we find w_1 is not

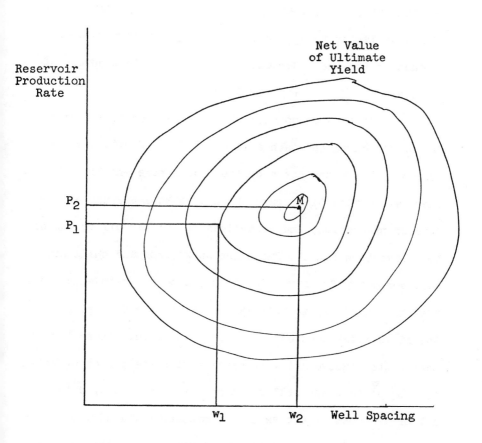

Figure 31 Net Value of Ultimate Yield Curves

the optimum rate of production or number of wells. More
present value of ultimate yield is possible by adding additional wells, up to w_2 where we have the optimum rate of
production for the reservoir P_2, and the optimum number of
wells w_2.

It is also possible to view this process from a
net profit curve approach as used by H. A. Scott. Here we
need to specify the costs of the system more exactly. First,
there are exploration costs which must be incurred regardless of production, and regardless of whether oil or gas or
both have been found. These are sunk costs and shall not be
considered any further. The second type of costs encountered
are those incurred in drilling, equipping, and developing
the field after the reservoir has been found. These investment costs involve and vary closely with the choice of well
spacing since these well spacing costs are a major determinant of the cost of access to the product. Thirdly, there
are those costs incurred in the maintenance of the production facilities, which will vary only slightly with the well
spacing chosen. Fourthly, there are those costs incurred
which are a function of the gross revenue rather than of the
spacing, such as leasing costs, royalty costs. Lastly, there
are tax payments and revenues which need to be included,

payments being a function of the total net income, and depreciation and depletion allowances adding substantially to income. These last two factors, depreciation and depletion allowances, will be postponed until we get to the programming aspects of the problem.

If the cost of drilling a well were zero, then the firm could drill enough wells on any given reservoir to achieve the highest rate of production consistent with physical ultimate recovery. The economics of the well spacing in this circumstance is uninteresting, and in such circumstances it is sufficient to construct only an ultimate physical recovery curve as in Figure 24 to determine the production rate, unconstrained by well spacing.

However, in fact wells are not a free good, and the individual firm must compare marginal returns with marginal costs for a different number of spacing patterns. In particular, the firm needs to consider the marginal net revenue gain from drilling a well against the marginal users cost of drilling a well. Where these two are equal will determine the optimum number of wells and the optimum rate of production (MER).

Assuming that a firm develops a reservoir as a unit, and that almost all costs are incurred at the beginning

of the development period, that is the only recurring costs
are the maintenance costs which are an insignificant proportion of total development costs, then it is possible to
derive discounted cost curve given in Figure 32 . The
different cost curves represent different well spacings and
all have been discounted at the same rate. They are shown
to be rising over time due to the lifting costs and the
maintenance costs which are incurred over the lifetime of
the reservoir. The optimal policy rule with regard to cost
of wells, would be to begin cautiously with wide spacing
pattern since it is always possible to increase the well
density over time as information becomes more perfect both
in an economical and physical sense. If we plot discounted
costs, well spacing and time, we derive Figure 33 which is
the discounted cost curve for the firm for alternative
spacing rules.

If now we superimpose the discounted gross revenue
curve and the discounted cost curves over time, we would
obtain over time the net profit curve. In Figure 34 , we
have diagramed the discounted revenue and the discounted
cost curves on the same graph and have obtained a series of
net profit curves. The net profit curves varies as the
discount rate varies, the higher net profit curves indicating

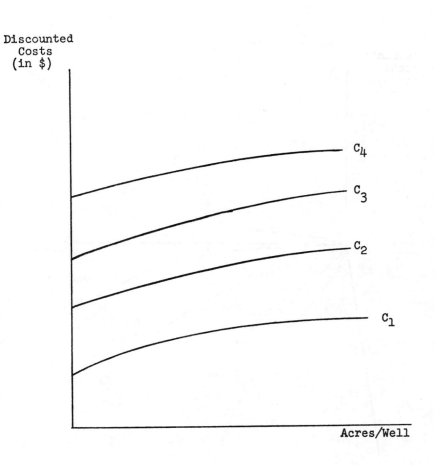

Figure 32   Discounted Costs and the Number of Wells

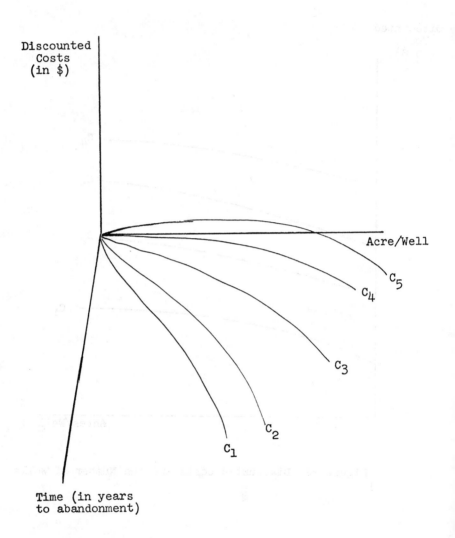

Figure 33   Cost Curves, Time, and Well Spacing
Patterns

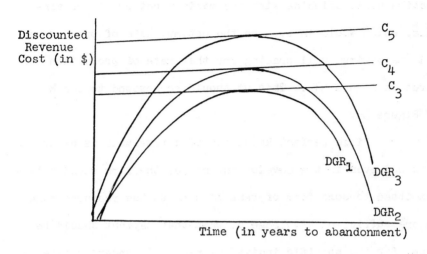

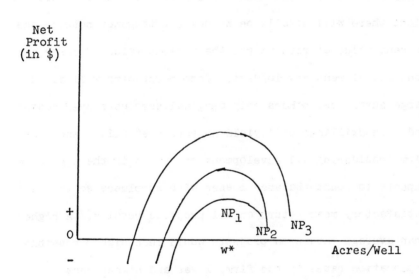

Figure 34   Net Profit Curves - Derivation

188

a lower discount rate has been used.[9] The point of production would coincide with the maximum net profit obtainable, which would in turn be the optimum rate of production
and the maximum well spacing for that rate of production
given in Figure 34 . This w* should correspond to the M
of Figure 35.

Since perfect knowledge of a reservoir is never
obtained even in the development stage, the firm usually is
committed to some form of rent payment to the resource owner
before sufficient knowledge of what that payment should be.
Thus, for the specific individual reservoir under consideration, there will usually be no perfect tangency between the
present value of returns and the present value of costs and
contractual rent development. Some reservoirs will yield
large surpluses, others only marginal surpluses over costs,
and some drillings will yield no return at all. Thus, the
firm considering all development prospects in the aggregate
expects to counterbalance losses with surpluses so that a
satisfactory mean return on all projects results. A higher
than average return on projects would mean higher possible
exploration rates by the firm, lower and higher rent

---

[9]See H. Anthony Scott's net profit analysis in
Chapter 2 of M. Gaffney (ed.), Extractive Resources and
Taxation (Madison, Wisconsin, Univ. of Wisconsin Press, 1968).

Reservoir
Production
Rate
STB/day

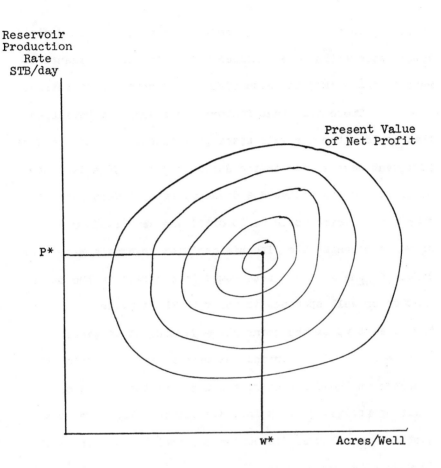

Present Value
of Net Profit

P*

w*   Acres/Well

Figure 35   Optimum Production and Well Spacing
with Net Profit Analysis

payments; lower than average returns on projects would yield lower exploration rates, higher prices and less payments to rent if the market is responsive to the behavior of return.

There are other factors and changes in variables that could alter the well spacing decision. One of the more important seen so far is the discount rate. If a firm had no profit motive and used a discount rate of zero percent, then its decision would be to drill the very minimum number of wells commensurate with his decision regarding reservoir life. As the discount rate comes into effect in the determination of well spacing, it tends to shift production into the present up to the point where discounted marginal revenue is equal to marginal users cost at which point the increase in the discount rate would have the effect of shifting production back into the future. Thus, we want just enough wells so discounted marginal revenue is equal to discounted users cost.

Another important variable is that of production and drilling costs, which are the major costs incurred by the firm in the development process. If costs are expected to increase (decrease) in the future, this would shift the slope of the cost curves in Figures 32 and 33 and shift the optimum depletion time to the present (future). A rise

(fall) in exploration costs for new discoveries would tend
to discourage (encourage) exploration and would cause firms
to discard (select) some areas previously considered
eligible for development. An expected rise (fall) in the
price of petroleum crude in the future would prolong the
rise (decline) of the discounted gross revenue curves in
existing reservoirs and would have the tendency of shifting
production toward the future (present), resulting in more
(less) intensive exploitation of deposits.

The same problem in more general mathematical form
to include all natural resources has been formulated by
V. L. Smith in a recent paper.[10] For the case of a nonreplenishable resources, gas and oil, he postulates the law
of growth to be $F(X) = 0$; that is, growth is impossible so
that over time the fixed stock for the industry can only
decline, and it declines at the rate of the industry output.

In its most general form, Smith defines the total
cost function for the individual firm to be of the general
form:

147
$$C = \phi\,(x,\ X,\ K) + \hat{\pi}$$

[10]V. L. Smith, "Economics of Production From
Natural Resources," American Economic Review, (June 1968),
Vol. 57, No. 3, Part I, pp. 409-431.

where $\emptyset$ (x, X, K) is fixed and variable operating cost, and $\hat{\pi}$ is the normal profit or return on a unit of capital required to hold the firm in the industry. The most natural general hypotheses about total operating costs for the firm in the oil industry Smith proposes are first, that operating costs are essentially independent of output so that $\partial c/\partial x = 0$; secondly, they are independent of the resource stock $c/X = 0$; and that they are independent of the real capital stock invested $\partial c/\partial K = 0$.

He then undertakes to characterize the competitive recovery process by a system of three behaviorial equations describing the interactions of the resource, individual firms, and the industry. If ρ (Kx) is the revenue derived from the sale of Kx units of the resources, and if output, x, of a given well can be regarded as a decreasing function of aggregate current output, Kx, and a decreasing function of cumulative output $\overline{X}$ - x,[11] then

148 $$x = g\ (Kx,\ X)$$

---

[11]This is so because costs are substantially independent of output so Smith proposes, and depends primarily on the characteristics of the field. Also both the rate of recovery and the cumulative recovery reduces the pressure level, so that output, x, can be regarded as a decreasing function of aggregate current output and cumulative output. See Smith, _Ibid._, p. 414.

$$\frac{\partial g}{\partial(Kx)} = g_1 < 0 \qquad \frac{\partial g}{\partial x} = g_2 > 0$$

New firms are assumed to be attracted into the
industry when normal profit is obtainable, that is, profit
defined by:

149 $$\pi = \frac{\rho(Kx)}{K} - c(x, X, K)$$

are such that $\pi > 0$. Specifically, Smith assumes that
the flow of capital into the industry is proportional to the
pure profits in the industry, or that:

150 $$\dot{K} = \partial \left[\frac{\rho(Kx)}{K} - c(x, K, K) \right]$$

where $\partial_1 \; \pi > 0$ if $\pi > 0$; and $\partial_2 \; \pi = 0$ if $\pi < 0$ which
means that in the case of wells, once drilled they cannot be
reduced in quantity.

Thus, Smith defines the fixed price equation system
for the petroleum industry as:

151 $$\dot{X} = - Kx$$
$$x = g(Kx, X) \quad g_1 < 0 \quad g_2 > 0$$
$$\dot{K} = \partial [R - C] \quad \partial_1 > 0 \text{ if } R > C$$
$$\partial_2 = 0 \text{ if } R < C$$

First, Smith observes that if it pays to drill any
wells at all, then at some point $K > 0$; and by equation
we know K can never fall. Thus, $\dot{X} = 0$ in equation 151 if

and only if $X = 0$, that is, the resource is not in equilibrium until well output is zero. Thus, the primary stage of recovery ends when $X = \underline{X}$ which is when $x = 0$ for the individual reservoir. Equilibrium in the industry is defined by $\dot{K} = 0$ or $c/\rho = g(Kc/\rho, X)$. Writing the total differential of this last equation and solving for dK/dX gives the slope of $I(X, K) = 0$:

$$152 \qquad \left. \frac{dK}{dX} \right|_{\dot{K}\,=\,0} = -\frac{\rho g_1}{c g_2} > 0$$

which says that the equilibrium size of the industry is an increasing function of the size of the untapped oil reservoir, X.

Assuming an initial system at $P_o(K = 0, X = \bar{X})$ that is, no capital invested, and the initial level of resource stock, then if $I(X, K) = 0$ is above p_o then it pays to exploit the resource or $\dot{K} > 0$, and hence $\dot{X} = -Kx < 0$ and the system will move along some path to a point ρ on $I(X, K) = 0$. Thereafter, $\dot{K} = 0$ but the stock declines, that is, $\dot{X} = -KX < 0$ until $x = 0$ at which point production stops unless secondary recovery is possible.[12]

Another formulation of the problem of optimal

---

[12]This is shown by diagrams by Smith on p. 424
Figure 7.

investment of the firm has been undertaken by Thompson and George.[13] Their presentation is concerned with using the calculus of variations to study dynamic economic problems, and specifically the dynamics of firm investment. Their general purpose is to maximize the discounted value of net profits from production less the costs of interest and investment in new capacity. The maximization is constrained by the change in capacity as well as debt. The reader is referred to the paper for a full development of the paper along the lines of control theory.

Lastly, at the present time, the Thompson-George analysis is being applied to the gas-oil industry by W. Kuller at the University of Kansas.[14] According to Professor El Horidi, the work is in the initial stages and no results are as yet reportable.

Actual conditions in the industry are quite different than have been discussed above. To consider the above problems in more realistic light, the ensuing chapters will discuss modifications of the analysis in this chapter,

---

[13]R. G. Thompson and M. P. George, "Optimal Operations and Investment of the Firm," Management Science, (Sept. 1968), Vol. 15, No. 1, pp. 58-64.

[14]The work is being supervised by Professor Ruppert and I thank him for bringing this work to my attention.

namely the determination of optimal production rates and

finally optimality of some profit function defined in

Chapter 8. Both linear and dynamic models are considered

since as the above analysis indicates, time and the time use

of resources is the essential core of the problem we are

considering.

CHAPTER 6

WELL SPACING AND PRODUCTION: LINEAR

AND DYNAMIC MODEL FORMULATIONS

Introduction

Well spacing, the rate of production, and ultimate

recovery are all interdependent variables. Thus, given a

specified ultimate recovery, the rate of production of the

reservoir varies directly with the number of wells, as was

seen in Chapter 5 on the economic considerations. A given

ultimate recovery can be obtained with very few wells if a

sufficiently long period of time is afforded for production,

or the same ultimate recovery can be obtained by drilling

more wells in a shorter period of time if needs be for a

shorter economic life. Similarly, given a specified pro-

duction rate for the reservoir, the ultimate recovery varies

directly with the number of wells drilled. Thus, we can

obtain the economic ultimate recovery curve and the dis-

counted gross revenue curves of Figures 24 and 25, and

notice that each one is really a family of curves that

depends on the particular well spacing chosen.

The physical relationship likewise is different for every reservoir, but as we have seen earlier, ultimate physical recovery is independent of the well spacing. However, the general form for the physical relationship can be seen from Figure 29 . The U curve or sets of U curves each represent a different ultimate recovery for a different rate of production and a different well spacing pattern. The production rate consistent with any given ultimate recovery can rise to a point as the wells are increased. It then remains constant reflecting the independence between well spacing and ultimate recovery. Increasing ultimate recovery is demonstrated by the shift of the U curve downward as more wells are drilled.

Within the above framework, we now want to consider the problem of petroleum production in the framework of linear programming and dynamic programming to see if rates of production and well spacing are determinable with the use of these techniques. It is safe to say that even the simplest reservoir problem is nonlinear in nature, and hence, the problem that is being studied needs to be considered in a nonlinear framework. While this is true, it is still possible and helpful to study a reservoir from a fixed geometric point of view, and to attempt to quantitize the time variable so that the resultant reservoir system may be described by a system of linear constraints and a linear objective function and thus obtain variable rates of production which in some way approximate reality.

General Model

The general form of the model which we shall use is as follows: first, we want to define the nature of the problem and the underlying assumptions made; secondly, the specifications of the decision variable(s) is needed for the analysis of the problem; thirdly, a specification of the constraining factors of the problem needs verbal and mathematical specification; and lastly, we shall consider the general and functional nature of the objective function used in the specification of the model.

The nature of the problem is to determine the rate of production of a reservoir with known ultimate recovery subject to certain restrictions which will maximize profits over the life of the reservoir. The specific restrictions and assumptions made will be with regard to the flow of fluids in an underground reservoir, the restrictions on surface facilities, and with regard to the optimum economic development of a reservoir, that is, maximizing net profit which is discounted over the life of the reservoir.

Specifically, the assumptions made are first that there exists a finite known amount of reserves in the given reservoirs calculable by the methodology described in Chapter 3. These reserves represent a homogeneous product, and any reservoir considered produces only a single product, either gas or oil not both. The demand for this homogeneous product is known to the firms, and it is assumed that the pipeline is the ultimate demander and consumer of the product. Likewise, there exists a finite array of possible production points in a reservoir that can be developed. Further, it is assumed that all wells drilled are homogeneous in their rate of production and flow characteristics. Lastly, production from the reservoir does not begin until all the necessary wells have been drilled, that is, the reservoir is completely developed.

As we have seen in Chapter 3, there exists an expression which represents the technology of the flow

---

[1]Linear models that have been developed similar to these will be found in: J. Aronofsky, A. Lee, "A Linear Programming Model for Scheduling the Production of Crude Oil," Trans AIME, (1958), Vol. 213, pp. 389-392. A. Charnes and W. W. Cooper, "Management Models and Industrial Applications of Linear Programming, Vol. II, (New York, John Wiley and Son Inc.,1961), Chapter 16, pp. 583-615.

characteristics of the reservoir. There are many characteristics which could be singled out, but the relationship that
we feel is most important is that which exists between well
pressure at any point in time and the rate of production.
For an oil reservoir the rate of production relationship is
given by the expression:

$$153 \qquad q_{sc} = \frac{7.08 \ kh \ (P_i - P_w)}{\mu B_o \left[\ln \frac{r_e}{r_w} - \frac{1}{2} \right]}$$

where everything has been defined as before in Chapter 4
Rewriting 153 above, we obtain expression 154 in terms of
well pressure:

$$154 \qquad P_i - P_w(t) = q_{sc} \frac{\mu B_o \ \ln(r_e/r_w - \frac{1}{2})}{7.08 \ kh}$$

Equation 154 expresses the initial state of the reservoir
before production begins, $(t=0)$. The initial pressure of
the reservoir is given by P_i, which is the shut-in pressure.
After development at time $t=0$ production begins at a constant flow rate, q_{sc}. Equation 154 then will give us at
any point in time $P_w(t)$, the flowing well pressure. It is
assumed that for each point in time the function:

$$155 \qquad F(t) = \frac{\mu B_o \ \ln(r_e/r_w - \frac{1}{2})}{7.08 \ kh}$$

has been calculated as the characteristics of the well

change under the flowing conditions so that for every point
in time F(t) is known, and thus the geometry of the reservoir is known. The time variable is also quantitized into
discrete time intervals t = 1 ... T. Thus the constraints
and the production rates are linear relations.

For the F(t) function -- commonly referred to as
the influence functions -- the various values for the permeability of the function over time are given and known, as
well as for the viscosity, the oil coefficient, and the
thickness of pay. It is also assumed that the initial well
pressure P_1 is known and that the external well radius, as
well as the well radius, is known and fixed for the reservoir.

Model I

The first case to be considered is that of
determining the rate of production from a single reservoir
with a single well for a series of time periods. While
this is not a very interesting case and there is little need
to place it in the linear programming framework, it has been
included here as the prototype of the models to follow. The
well pressure at any point in time then can be given by a
rearrangement of Equation 155:

156 $$\sum_{t=1}^{T} P_W(t) = P_o - \sum_{t=1}^{T} \left[q(t) - q(t-1) \right] \cdot F(t-t-1)$$

where:

$$\sum_{t=1}^{T} q(t) - q(t-1)$$ the average rate of production in the tth time period

$$\sum_{t=1}^{T} F(t-t-1)$$ the change in the reservoir parameters over the tth time period;

where well pressure $P_w(t)$ at any point in time can take on any value within bounds of course. The bounds are P_i the initial well pressure and P_a the abandonment pressure. Thus, the first restriction can be expressed:

157 $$\sum_{t=1}^{T} P_W(t) = P_a.$$

The next restriction is one imposed by the reserves of the reservoir. It states that the cumulative physical recovery cannot exceed the initial amount of reserves in the reservoir so that:

158 $$\sum_{t=1}^{T} q(t) - q(t-1) = R_p$$

where R_p, the physical maximum recovery, is given by the materials balance equation and is not actual total reserves

but maximum reserves that are physically recoverable by
present techniques. Thus, this physical maximum recovery
could change as the technology changes over time, or as the
characteristics of the reservoir become known over time.

The last restriction imposed is of the form that
the production rate during any time period (t-t-1) cannot
exceed the above ground facilities or the pipeline capacity
at the given reservoir:

159
$$\sum_{t=1}^{T} q(t) - q(t-1) = L_c$$

where we define L_c to be the pipeline capacity for the
reservoir.

All the restrictions of this model are given by
Equations 157, 158, 159. We have T constraints in the unknown
variable q where q is over time $\sum_{t=1}^{T}$. The problem now is to
find from the set of all values of q satisfying the constraints, that rate of all production which will maximize
profit over the lifetime of the reservoir:

160
$$\text{Max } \pi\big[q(\ t)\big] = R\big[q(\Delta t)\big] - c\big[q(\Delta t)\big]$$

The mathematical structure of the model depends on
the assumptions made about R and C as a function of the
decision variable q.

In traditional economic theory there are several

alternative forms this objective function can take, and the most common ones are diagramed in Figure 36 . In words and in mathematical symbols these are expressed as follows:

$$R = pq \quad \text{(linear revenue)}$$
$$R = pq^{\alpha} \quad \text{(concave revenue)}$$
$$C = a+bq \quad \text{(linear cost)}$$
$$C = a+bq^{B} \quad \text{(convex cost)}$$

or linear revenue, linear cost; linear revenue, convex cost; concave revenue, linear cost; concave revenue, convex cost.

The terms in the objective function, $R\left[q(\Delta t)\right] - C\left[q(\Delta t)\right]$, are the total profit over the entire lifetime of the project.

If both revenue and costs are linear, then the analysis becomes that of the classical breakeven analysis. If revenues are given by $R_1 = p\left[q(\Delta t)\right]$, and if costs are of the form:

161
$$C = FC + FO + b\left[q(\ t)\right] + S\left[q(\Delta t)\right]$$

where:

p = price;

FC = fixed cost;

FO = fixed operating cost;

b = variable costs;

S = variable operating costs;

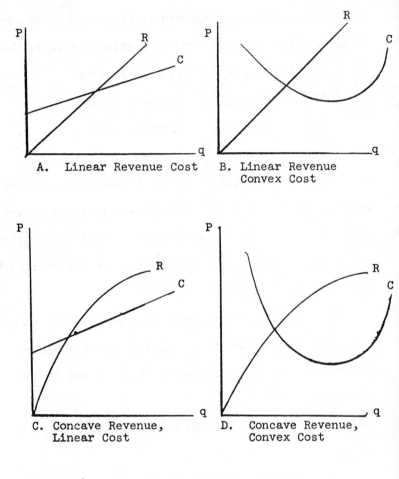

A. Linear Revenue Cost

B. Linear Revenue Convex Cost

C. Concave Revenue, Linear Cost

D. Concave Revenue, Convex Cost

Figure 36   Forms of the Objective Function - Diagramatic

then the earnings function can be reduced to the simple form:

162
$$E = \alpha \left[q(t) - q(t-1) \right] - FT$$

where:

α = variable earnings factor;

FT = FC - FO = total fixed cost.

The breakeven point then is given in Figure 37 by XA. When other forms of the revenue and cost functions are being considered, the earnings function becomes concave with a unique maximum. These other cases are aggregated in Figure 37. The form of the profits function or earnings function will be examined in more detail at a later point when we consider depletion allowance and depreciation allowance in the earnings function.

This model developed is a comparatively simple one and really does not need linear programming as a tool for solution. It is possible to solve a one well one reservoir-problem by simple manipulation. In the real world this model has little applicability seeing that it would be an unusually small well and thus doubtful if the reservoir should be considered for production at all. However, it is the structural process which concerns us and the implications that the model may have for the more complicated process and models later considered which justifies the

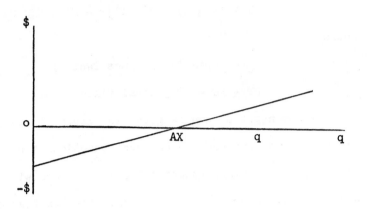

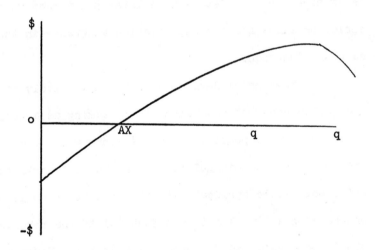

Figure 37 Breakeven Curves

consideration of Model I. In the next section, we shall
then be concerned with finding the rate of production for
more complicated problems and from this infer well spacing.

The above model was for an oil reservoir only.
If we consider a gas reservoir under the same assumptions
and conditions, the reservoir behavior problem is not linear
in space, and thus even though we quantitize the time variable, a nonlinear program results. This can be seen by
reviewing Equation 163, the gas production equation:

163
$$q_{sc} = \frac{703 \ kh(P_e^2 - P_w^2)}{\mu TZ \ \ln(r_e/r_w)}$$

Then developing the relation between well pressure and the
production rate yields us:

164
$$P_e^2 - P_w^2 = q_{sc} \frac{\mu TZ \ \ln(r_e/r_w)}{703 \ kh}$$

It is convenient again to allow the influence function to be
given by:

165
$$P(t) = \frac{\mu TZ \ \ln(r_e/r_w)}{703 \ kh}$$

so that 165 becomes:

166
$$P_e^2 - P_w^2(t) = q_{sc}(t) \ F(T) \ .$$

But we are still left with the problem of P_e^2 and $P_w^2(t)$.
In order to solve for the unique set of values of q, we would

have to indulge in nonlinear programming techniques which for the moment it seems better to ignore. Thus, we shall have to satisfy ourselves with only the oil situation, and realize that the gas problem has similar characteristics, form, and solutions of a more complex nature.

Model II

The second model that will be considered is more general than the first, and probably a little more interesting. A not uncommon experience in drilling reservoirs is to find faults which really divide the large reservoir pool into separate single reservoirs with no interference between the separate pools of the reservoir. This separation of the reservoir is also very common when the reservoir rests on top of a salt dome; or certain geological structures tend to cause this separation of a given reservoir into pools. In all of these circumstances where there is no pressure interference -- for whatever reason -- it is possible to treat the reservoir as separate pools or separate reservoirs.

The problem in this case then becomes to compute the production rate from these sources -- two or more -- which over time and subject to the limiting conditions will again maximize profit.

The format of the model is similar to the first

developed. For each pool of the reservoir there is an
expression which represents the technological flow of a
fluid through a porous media relating the well pressure to
the production rate. Again for oil only, this is given by
Equation 153. Rewriting this equation, we again establish
Equation 154 and again define 155 as the influence function
for a specific pool of the reservoir. F(t) -- the influence
function -- will now be separately defined for every pool in
the reservoir. The problem now is to find the rate of production for many pools in a reservoir for a series of time
periods such that the restrictions imposed are satisfied
subject to profit being a maximum.

Variations in the flow rates are again obtainable
by using Equation 156 but every pool j must be specified so
that Equation 156 now becomes:

167
$$\sum_{t=1}^{T} \sum_{j=1}^{N} P_{wj}(t) = \sum_{j=1}^{N} P_{oj} - \sum_{t=1}^{T} \sum_{j=1}^{N} q_j(t)$$
$$- q_j(t-1)F_j(t-t-1)$$

Again, the well flow pressure $P_{wj}(t)$ can take on any value
as long as in any pool j it does not go below some abandonment pressure, where every pool has its own P_{aj}. Thus our
first restriction is of the form:

168
$$\sum_{t=1}^{T} \sum_{j=1}^{N} P_{wj}(t) \geq P_{aj}.$$

The next set of restrictions are the same as before only
the recoverable reserves for each pool are calculated by the
materials balance equation and we require that cumulative
recoverable production from each pool cannot be greater than
the reserves of that pool so that:

169
$$\sum_{t=1}^{T} \sum_{j=1}^{N} q_j(t) - q_j(t-1) \leq \sum_{j=1}^{N} R_{pj}$$

where R_{pj} is now the physical recoverable reserves from pool
j.

Lastly, we have the restriction that the production in any time period from any pool in the reservoir
system cannot exceed the capacity of the gathering lines
from pool j, and that the sum of the capacity of the gathering lines cannot exceed the trunk line capacity. These
restrictions can be written in the form:

170
$$\sum_{t=1}^{T} \sum_{j=1}^{N} q_j(t) - q_j(t-1) \leq \sum_{j=1}^{N} L_{cj}$$

171
$$\sum_{t=1}^{T} \sum_{j=1}^{N} q_j(t) - q_j(t-1) \leq W_c$$

where W_c is now the trunk line capacity.

Adding the constraints together we have TN - N - T in the unknown variables $q_j(t)$. The problem then becomes to find from all the sets of values which $q_j(t)$ can take, and which satisfy Equations 167, 168, 169, and 170, that set of values for which profit will be a maximum, that is, that set of values for which the objective function will be a maximum:

$$172 \qquad \text{Max} \qquad q_j(t) - q(t-1) = R \ q_j(t)$$
$$- C \ q_j(t)$$

Thus, we have basically the same problem that we saw in Model I, only the dimensions of the problem have expanded and thus the problem is not as obvious nor is the answer as obvious as it was in the previous case. An answer is obtainable by using the Simplex Algorithm to the problem as it is here formulated. The reader is referred to Charnes and Cooper for solutions to this particular model.[2]

Model III

The third production model that needs to be considered is when there is the possibility of interference between wells. The model considered here is for a single

---

[2]See Charnes and Cooper, _Ibid._, pp. 583-615.

reservoir with more than one well, and thus we need to consider the possibility of well interference and its effect on
the rate of production.

Assume first that there exists a single reservoir
with only two wells. Number the wells a and b, and these
wells are separated by a distance d. Then the relationship
between well pressure and the rate of production in time
period t for well a assuming interference can be written:

173
$$P_{wa}(t) = P_{oa} - \left\{ [q(t) - q(t-1)]_a \; F_a(t-t-1) \right.$$
$$\left. + [q(t) - q(t-1)] \; F_b(t-t-1)r_{ab} \right\}$$

where we now define:

$F_a(t-t-1)$ as the influence function for well

a in time period t;

$F_b(t-t-1)$ as the influence function for well

b in time period t;

r_{ab} as the well spacing ratio, the distance

between two well centers, that is, the

well radius.

For well b then the relationship will be given by:

174
$$P_{wb}(t) = P_{ab} - \left\{ [q(t)-q(t-1)]_b \; F_b(t-t-1) \right.$$
$$\left. - [q(t)-q(t-1)]_a \; F_a(t-t-1)r_{ab} \right\}$$

Generalizing to a multi-well system or reservoir
then it is possible to write Equations 173 and 174 in more

general form indicating the relationship between well flow
pressure at the 1th well and the rate of production for the
1th well:

175
$$\sum_{\ell=1}^{L} \sum_{t=1}^{T} P_{w_\ell}(t) = \sum_{t=1}^{T} P_o - \left\{ \sum_{\ell=1}^{L} \left[\sum_{t=1}^{T} q(t) - \sum_{t=1}^{T} q(t-1) \right]_\ell \sum_{t=1}^{T} \sum_{\ell=1}^{L} F(t-t-1)r_e \right\}$$

The constraints for this new problem are again of
similar form as before only now we allow for ℓ wells so that
the pressure constraint can be written:

176
$$\sum_{t=1}^{T} \sum_{\ell=1}^{L} P_w(t) \geq \sum_{\ell=1}^{L} P_{a\ell}$$

which says again that the well pressure in the 1th well
cannot go below the abandonment well pressure in that 1
well.

Secondly, the quantity produced from these wells
cannot exceed the amount of recoverable reserves from the
reservoir so that the constraint takes the form:

177
$$\sum_{t=1}^{T} \sum_{\ell=1}^{L} \left[q(t) - q(t-1) \right]_\ell \leq R_p$$

Further, no more can be produced from any single

well than the pipeline gathering system from the 1th well
will permit, and the total of the gathering system must be
equal to or less than the capacity of the trunk line. These
two restrictions can be expressed again by:

$$178 \qquad \sum_{t=1}^{T} \sum_{\ell=1}^{L} \left[q(t) - q(t-1) \right]_{\ell} \leq \sum_{\ell=1}^{L} L_{cl}$$

for the gathering system and

$$179 \qquad \sum_{t=1}^{T} \sum_{\ell=1}^{L} \left[q(t) - q(t-1) \right]_{\ell} \leq W_c$$

for the trunk line capacity.

The restrictions on this problem are LT - L - T
and the unknown are again the $q_{\ell}(t)$. The problem is again
to maximize Equation 180 , the objective function, subject to
the constrainting Equations 176, 177, 178, and 179:

$$180 \qquad \text{Max } \mathbb{\pi} \left[q(t) - q(t-1) \right]_{\ell} = \sum_{t=1}^{T} \sum_{\ell=1}^{L} R \left[q(t) - q(t-1) \right]_{\lambda} - C \left[q(t) - q(t-1) \right]_{\ell}$$

Model IV

The last linear programming formulation of the
models that we are working with considers the case where
there are multi-reservoirs for the reasons mentioned before.
Now we wish to extend Model III to the case where we

consider multi-wells and multi-reservoirs. These reservoirs are such that it is feasible to drill more than one well, and thus we are able to consider the multi-well, multi-reservoir case. The case is very similar to the one considered before, only the number of variables will change to allow for the change in the number of reservoirs.

Again, we start with the equation expressing variations in the flow rate q to the flow pressure which is now given in Equation 181:

$$181 \qquad \sum_{t=1}^{T} \sum_{j=1}^{N} \sum_{\ell=1}^{L} P_{wj\ell}(t) = \sum_{j=1}^{N} \sum_{\ell=1}^{L} P_{o,j,\ell} - \left\{ \sum_{t=1}^{T} \right.$$

$$\left. \sum_{\ell=1}^{N} \sum_{\ell=1}^{L} \left[q(t) - q(t-1) \right]_{\ell j} F_{\ell j}(t-t-1) r_e \right\}$$

where we assume there can be influence between the 1th wells, but no influence between the j pools.

The constraints for the new problem are again similar to the constraints of the previous problems, only the dimensions are greater. For well pressure, the constraint is of the form:

$$182 \qquad \sum_{t=1}^{T} \sum_{j=1}^{N} \sum_{\ell=1}^{L} P_{wj\ell}(t) \leq P_{aj\ell}$$

Now the quantity produced from the 1th well in the jth pool

cannot exceed the recoverable reserves of the pool or of the reservoir so that the reserves constraint is:

183
$$\sum_{t=1}^{T} \sum_{j=1}^{N} \sum_{\ell=1}^{L} \left[q(t) - q(t-1) \right]_{j\ell} \leq R_{cj}$$

Again every well has its own pipeline gathering system and each pool has its trunk pipeline, and the entire set has a major transportation line which cannot be exceeded at any point in the link. These constraints can be expressed by Equations 184, 185, and 186:

184
$$\sum_{t=1}^{T} \sum_{\ell=1}^{L} \left[q(t) - q(t-1) \right]_{\ell} \leq L_{c\ell}$$

185
$$\sum_{t=1}^{T} \sum_{j=1}^{N} \sum_{\ell=1}^{L} \left[q(t) - q(t-1) \right]_{j\ell} \leq W_{j\ell}$$

186
$$\sum_{t=1}^{T} \sum_{j=1}^{N} \sum_{\ell=1}^{L} \left[q(t) - q(t-1) \right]_{\ell} \leq T_{\ell j}$$

where now $T_{\ell j}$ is the major transformation line capacity.

The restrictions to this problem are now L T N-T N-T-N more than we have seen previously in the other models. The objective is again to find the set of values for $q_{j\ell}(t)$ such that constraints 181-186 are satisfied and such that again the profit function is maximized:

187   Max $\pi \left[q(t) - q(t-1) \right]_{j\ell} = R \left[q(t) - q(t-1) \right]_{j\ell}$
              $- C \left[q(t) - q(t-1) \right]_{j\ell}$

where the profit is again the dollar profit over the lifetime of the project.

What we have so far is the set of optimal average
production rates over time for the various assumptions made
in regard to the existence of wells and reservoirs. What
we are interested in doing is going from the optimal average
production to the well spacing for a given reservoir. This
is not difficult since once we have obtained a q*, an optimal
production rate, dividing this q* into the percentage of
emulative recoverable reserves gives us the number of wells
which can and should be drilled for each reservoir. The
number of wells, say l, times the cost per well, assuming
all well costs are constant for a given reservoir, will
yield the investment cost for drilling a given reservoir.
This in a simplified sense gives us the optimal well spacing
and the investment cost for a given reservoir.

As for where the wells should be placed, this is a
problem in geometry and geology and has to be solved
according to the specific characteristics of flow for the
specific reservoir. We in this paper are not concerned with
the exact location, but rather with the investment cost and

the number of wells necessary to optimally drain the given
reservoirs.

Before going into the problem of investment models,
we would like to spend the next few pages developing a dynamic model for the determination of optimal production rates,
and thus achieve a further extension of the linear models
here considered.

Model V

The problem which now concerns us is putting the
linear models into a dynamic framework so that given an
array of optimal production rates, it is possible to maximize the profit function as defined herein. Before considering the optimal profit function, which now lets us
classify as the main problem, we want to consider the sub-
problem of determining the optimal rates of production over
time. In order to consider this problem, we need to define
the following variables and functions:

t = the index of periods which starts with zero

for the beginning of the development period

and ends with time T;

i = the index of wells and facilities. We assume

the existence of well i which we call the

experimental well, and it is possible to add

2.....n wells in the following period;

P_{o1} = the initial flow pressure at time t = 0 at well i;

P_{wit} = the well flow pressure at any time at the ith well;

P_{ait} = the well abandonment pressure at time t at the ith well;

P_{itmin} = minimum allowable flow pressure at time t at well i.

P_{itmax} = maximum allowable pressure drop above which will damage or decrease ultimate physical production;

k_{it} = permeability in well i at time t;

μ_{it} = viscosity in well i at time t;

h_{it} = net pay thickness in well i at time t;

B_{oit} = oil coefficient in well i at time t.

The first sub-problem then can be defined as follows:

188
$$\underset{P_{wit}}{\text{Max}} \quad q_{sct}$$

subject to:

189
$$P_{it\ min} \leq P_{wit} \leq P_{it\ max}$$

$$P_{wit} \gtrless P_{ai}$$

The q_{sct} can be defined as before only now it can vary over time and thus we time subscript the variables:

190
$$q_{sct} = \frac{7.08 \, k_{it} \, h_{it} \, (P_{eit}-P_{wit})}{\mu \, B_{1to} \, [\ln r_{eit}/r_{wit}-\frac{1}{2}]}$$

The mathematical structure of the model depends on the assumptions made about k, μ, h, and B_o as functions of the decision variable B_{wit}, well flow pressure. k, μ, h, and B_o, of course, depend on the type of drive used and the reservoir in question. These variables are physical variables and their effects on production need not be repeated here since they have been considered in some detail in Chapters 2 and 3.

The second sub-problem that needs to be considered is that once given these optimal q_{1t}^*, we want to maximize the profit function subject to these q*'s over time. For this problem we need to now define the following variables:

D_t = forecasted demand for time t for the given reservoir;

q_{1t}^* = optimal production in facility i in time period t.

q_{1tmin} = minimum productive capacity at facility i for time period t;

q_{1tmax} = maximum productive capacity at facility 1
for time period t;

R_1 = Revenue as a function of q_{1t}^*;

C_1 = Cost as a function of q_{1t}^*.

It is now possible to define the second sub-
problem as:

191
$$\text{Max } \pi_t$$
$$q_{1t}$$

subject to:

192
$$q_{1t \text{ min}} \leq q_{1t}^* \leq q_{1t \text{ max}}$$

and

193
$$q_{1t}^* \leq D_t$$

Here we defined π_t as net profit discounted over time or:

194
$$\pi_t = R_1(q_{1t}^*) - C_1(q_{1t}^*)$$

Again the mathematical structure of the model depends on the
assumptions made about R_1 and C_1 as functions of the decision variable q_{1t}. Here for simplicity sake, we assume that
both are strictly linear functions, which allows us to
ignore many of the complicating aspects of the problem. The
profit function above, Equation 194, will be modified sub-
sequently in Chapter 8 to include depletion and depreciation
allowances. However, here we are only trying to view and

understand the productive aspects of the problem. The profit function will be considered further in Chapter 8, along with debt and equity constraints.

The solutions to these sub-problems now formulated give us for each period the optimal rate of production and the maximum profit attainable at that rate of production. Assuming that the solutions to the sub-problem exist, it is then possible to formulate the mathematical structure of the main problem.

There exists a decision variable, let us call it W_{sit}, for each possible well spacing alternative such that:

195
$$W_{sit} = \begin{cases} 1 \text{ if a well is drilled in time period t} \\ 0 \text{ if a well is not drilled in time} \\ \text{period t.} \end{cases}$$

Also, for each alternative W_{sit}, there is a set of mutually exclusive variables such that if a well is drilled in time period $T = t$, it cannot be drilled in any other time period. Thus $0 \leq W_{sit} \leq 1$. It should be noted that the index ranges over i, not only the set of possible alternatives, but also over the set of existing wells so that at $t = 0$, we find that $W_{si0} = 1$; that is, at time period zero at least one well has been drilled, let us call it the experimental well.

225

Another variable ξ_t is also defined, where ξ_t is determined by W_{sit}. It is defined as a vector having as many elements as there are periods in the planning horizon, and ξ_t has zeros before W_{sit} is initiated and ones thereafter. For example, some of the ξ_t's would look as follows:

$$\xi_1 = [1, 1, \ldots 1]$$

196

$$\xi_2 = [0, 1, 1, \ldots 1]$$

$$\xi_3 = [0, 0, \ldots 0, 1]$$

$$\xi_n = [0, 0, \ldots 0, 1]$$

This variable is used with W_{sit} and its function is to indicate when another should be drilled, and thus when it is possible to increase profits by drilling another well.

Then for any given time period t, and a given set of possible well alternatives W_{sit}, it is possible to define Equation 197:

197

$$\sum_{i=1}^{n} \sum_{t=1}^{T} \xi_t \, W_{sit} \, \pi_t$$

which indicates the optimal profit obtainable via optimality in the various sub-problems. Discounting this profit function over time by the appropriate discount rate α

where α is the discount factor over time t, and thus Equation 197 yields us the present value of the sum of the profits over time and over the planning horizon for each period.

The objective function of the main problem can then be defined as:

198
$$\underset{W_{sit}}{\text{Max}} \sum^{T} \alpha^t \left[\sum^{n} \varepsilon_t \; W_{sit} \right] \pi_t$$

and this problem will be maximized by the proper choice of W_{sit}.

Since the drilling of wells costs money, there is associated with each well a cost parameter. This parameter has as many elements as there are periods in the planning horizon. It is defined as having a non-zero element starting in the tth period, and extends for as many periods as the drilling costs require. Hence, if drilling the ith well in a reservoir requires C_1, C_2, C_3, costs, then the C_i function would have the form:

199
$$C_i = (0, \ldots C \; , \; C \; , \; C \; , \; 0, \ldots \ldots 0)$$

It is assumed that the ith well is capable of production in the last period in which the costs are incurred, that is, period C_{i-3}.

The remaining constraints imposed in the formulation of the dynamic model are the physical limitations of a
similar form as were imposed in the linear models. However,
reformulation of the constraints is necessary for the dynamic model, and dynamic consideration. The necessary
restraint on reserves at any time period t is that:

200 $\qquad R_t = R_o - q_{scit} \, t \, W_{si}$

where:

R_t = the reserves available at time period t;

R_o = the initial reserves that are physically
recoverable;

q_{scit} = the production rate under standard operating
procedures for the ith well in the tth time
period;

t = the number of time periods;

W_{si} = the number of wells on reservoir i.

For the calculation of the initial reserves R_o, the reader
is referred to Chapter 3. A further necessary condition is
that for any time period t:

201 $\qquad q_{scit} \leq L_{ci}$

which says that the production from any well i from the
reservoir cannot exceed its gathering line or its pipeline
capacity. Further, we have the restriction that:

202 $\qquad q_{scit} W_{sit} \leq W_{1t}$

or that the production from the entire reservoir at any time
period cannot be more than the trunk line capacity for any
time period under consideration. The pressure constraints
have been already considered in the analysis of the first
sub-problem, and thus need no further consideration here in
the restraint section.

The optimization technique applicable in all cases
is that of dynamic programming with each well being a stage
in the decision-making process. The state variable is the
amount of production that is being allocated. If we let:

q_{sc} = the amount of production being allocated;

$q_{1,max}$ = the maximum capacity at reservoir i;

$q_{1,min}$ = the minimum capacity at reservoir i;

N = the number of possible wells under
consideration;

$\pi_1(q_{sc})$ = the profit at reservoir i when q_{sc} is
being allocated;

then several restrictions apply before representing the
problem as a series; namely:

203 $\qquad q_{1,min} \leq q_{sc1} \leq q_{1,max}$

204 $\qquad A_1 = Max \quad q_{sc} - \sum_{j=1}^{i-1} q_{j,min}$

is the lower bound on production at reservoir i:

205 $B_i = \text{Min} \quad q_{sc} - \sum_{j=1}^{i-1} q_{j,max}$

is the upper bound on production at reservoir i:

and;

$$F_0(q_{sc}) = 0.$$

Then we can formulate the problem as a series in Equation 206:

$$F_1(q_{sc}) = \max_{A_1 \le q_1 \le B_1} \left\{ \pi_1(q_1) \right\}$$

206

$$F_2(q_{sc}) = \max_{A_2 \le q_2 \le B_2} \left\{ \pi_2(q_2) - F_1(q_{sc} \quad q_2) \right\}$$

$$F_i(q_{sc}) = \max_{A_i \le q_i \le B_i} \left\{ \pi_i(q_i) - F_{i-1} (q_{sc} \quad q_i) \right\}$$

$$F_N(q_{sc}) = \max_{A_N \le q_N \le B_N} \left\{ \pi_N(q_N) - F_{N-1} (q_{sc} \quad q_N) \right\}$$

The algorithm used to solve this problem is based on Bellman's principle of optimality and is covered in greater detail in Chapter 9.[3] Also, results have been simulated to show that the model is workable and can be used

---

[3]Chapter 9 will present a dynamic programming algorithm which will then be used to solve the production and investment problem as it has been formulated here.

on the optimal operation of a reservoir, that is, within
the restrictions imposed.

CHAPTER 7

CAPITAL INVESTMENT THEORY IN

THE LITERATURE: A REVIEW

Introduction

The problem of capital investment has engaged

economists for some time. Prior to World War II, the econo-

mist focused primarily on the fundamental issue of the nature

of the rate of interest.[1] Implicit in their treatment of

the rate of interest was a model of optimal investment for

the firm. Like most of the economic theory of the time,

their models were simplified, making use of assumptions

regarding the continuity and differentiability of the func-

tions they treated or postulated, as well as using models of

certainty within the structure of a perfect capital market.

The key proposition of the models they developed was that an

investment program would be carried on by the firm to the

---

[1]See K. Wicksell, Lectures on Political Economy,
(New York, 1934)

point at which the marginal rate of return to the firm was equal to the market rate of interest.[2]

While these models might today be considered crude, they did handle an actual management problem, that of investment planning in relation to time.

Based on the economic theory of investment, Joel Dean presented a practical working alternative to the classical investment procedure, namely capital budgeting.[3] Reduced to its simplest terms, Dean's idea is for the individual firm to compute an internal rate of return for each investment alternative and then to rank these investment projects in decreasing order according to this internal rate of return criterion. The intersection then of the internal rate of return with the marginal cost of capital schedule determined the cut-off point, or the region of minimum acceptable rates of return which then could be used for selecting various investment alternatives. Thus, for the individual firm the investment process can be viewed as in

[2]R. D. G. Allen, Mathematical Analysis for Economist, (London, The MacMillan Co., 1938), pp. 374-378.

[3]Joel Dean, Capital Budgeting, (New York, John Wiley and Son, Inc., 1951). Also in his work, Managerial Economics, (Englewood Cliffs, N.J., Prentice Hall, 1951) are contained many of the basic concepts presented in his Capital Budgeting treatise.

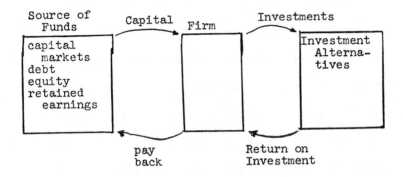

Figure 38   Capital Investment Decision

Before reviewing the direction taken by the

economist and the financial analyst, it is important to

spend some time reviewing a general presentation of capital

investment problem.

The Capital Investment Problem

A capital investment may be defined as an expendi-

ture by a firm which is designed to provide benefits in

future periods. The distinction between capital investment

expenditures and other expenditures by the firm is for

income tax accounting purposes. The capital investment

expenditure cannot, in an accounting sense, be treated as

an expense in the period in which the expense is incurred,

but rather is depreciated over its useful life. Thus, with

each capital investment is associated: (1) the life of the

investment; (2) the time stream of costs; (3) the time

stream of revenues. The firm can invest in investment projects from which it receives benefits. In order to invest
it needs to use capital either in the form of retained
earnings, or it needs to borrow, in which case it pays an
interest charge on the funds borrowed. Also, it may in
future periods use the surplus of funds coming from past
investments to invest in new investment alternatives available to the firm.

Traditionally, the present value concept has been
used for analyzing investment problems. The present value
of any investment then can be given as a function of the
discount rate:

207
$$PV_i(\alpha) = \sum_{t=1}^{m} \frac{a_{it} - b_{it}}{(1 + \alpha)^t}$$

where:

PV_i = the present value of investment i;

a_{it} = revenues received from investment i in
t periods;

b_{it} = investment costs incurred for investment
i in t time periods;

α = discount rate.

The traditional rule is that as long as $PV_i > 0$, the investment should be accepted. However, there is a difficulty
with this rule, namely α is assumed to be the cost of

capital to the firm, i.e., the interest rate the firm has to pay when it borrows money. Clearly, the total present value of investment alternatives to the firm will be greater if a project with positive present value is added than if it were not added.

Financial analysts have found it difficult to define and measure a firm's cost of capital. The general consensus is that there is no single cost of capital for any firm, but really the cost of capital depends on an entire array of factors.[4] Other theorists have advocated the use of a decision rule based on the internal rate of return for the individual firm.[5]

The internal rate of return can be defined as that rate which equates the stream of costs incurred to the stream of revenues received. In terms of the net present values, the internal rate of return α is that rate for which the net present value of an investment is equal to zero, i.e.

208    $$PV_1 = \sum_{t=1}^{m} \frac{a_{1t} - b_{1t}}{(1 + \alpha)^t} = 0$$

[4]See W. Lewelyn, The Cost of Capital, (Belmont, California, Wadsworth, Inc., 1968).

[5]See Joel Dean, Capital Budgeting, (New York, John Wiley and Son, Inc., 1951), as one of the first presentations of this rule.

The decision rule in this case is that the investment is accepted as long as $\alpha > \alpha$ min. If $\alpha \leq \alpha$ min, then the investment is rejected since $PV_1 < 0$. This rule leads to the same result as before and like the previous method is acceptable for a single investment alternative.

A natural extension of these decision rules is to rank the mutually exclusive set of alternatives according to their rate of internal return, or according to the net present value contribution. Where there is a limit on the amount of funds available, the various alternatives are ranked according to one of the above methods and the investment alternatives then selected according to rank until all the funds available have been used. This procedure, however, could lead to a non-optimal solution, if only because of the discrete size of investment projects left some funds unused. It is possible to obtain an optimal distribution of funds via dynamic programming.[6]

The use of these two methods of ranking, internal rate of return and net present values, can lead to conflicting rankings. This can arise because of mixed projects,

[6]See W. J. Fabrycky and P. E. Torgersen, Operations Economy: Industrial Applications of Operations Research, (Englewood Cliffs, N.J., Prentice Hall, 1966), pp. 426-429.

i.e., some projects may entail both lending and borrowing, and both methods make certain implicit assumptions concerning the reinvestment of cash flows from the project.[7]

Thus, in this chapter the problem of interest can be identified as: (1) a firm has a single investment possibility and it must decide whether to accept or reject it; (2) a firm has a number of possible investment projects and it may accept or reject one or more; (3) both (1) and (2) are limited by the amount of funds that are available for the firm's investment process.

Classical Approach

Post-war research in capital investment theory has taken two distinct directions; one which we shall term classical, the other we might term programming or techniques which involve the use of modern programming theory and techniques. Both have in common the approach of optimizing some objective function subject to a set of constraints; only the programming approach makes use of this technique

[7]For an extended analysis of these problems the reader is referred to: E. Soloman, "Measuring a Company's Cost of Capital," Journal of Business, (Oct., 1955); J.C.T. Mao, "The Internal Rate of Return as a Ranking Criterion," The Engineering Economist, (July-Aug., 1966); R. H. Bernhard, "Discount Methods for Expenditure Evaluation," Journal of Industrial Engineering, (Jan.-Feb., 1962), pp. 19-27.

238

more explicitly than the classical approach.

The first school we would like to briefly discuss
is that of the classical economist. In the pre-World War
II literature, Irving Fisher exposed his views on the
theory of capital in two works cited below.[8] Fisher showed
that equilibrium under capital rationing requires the
investor's subjective discount rate between two periods to
be equal to the marginal rate of transformation between outputs in those two periods.[9] That is, the discount rate must
equal the ratio of the marginal yields of the investor's
resources in those periods. This result basically means
that in equilibrium, the investor will be on the frontier
of the production investment possibility set.

In a more recent article, Hirshleifer (1958)
attempted to apply the Fisherian analysis through the use
of isoquants to the problem of optimal investment decision
making.[10] In the article, Hirshleifer arrived at the same

---

[8]Irving Fisher, The Theory of Interest, (New York,
The MacMillan Co., 1930), and The Rate of Interest, (New
York, The MacMillan Co., 1907).

[9]See Fisher, (1930), Ibid., pp. 266-269.

[10]J. Hirshleifer, "On the Theory of Optimal
Investment Decision," Journal of Political Economy, (1958),
Vol. 66, pp. 329-352.

rule Fisher did in his analysis, but extended the result to a multi-period investment process.

Another alternative treatment of capital theory is implicit in the Walras-Wicksteed marginal productivity theory of production. The treatment therein is itself a short-run theory of operating techniques for individual firms. However, the Walras-Wicksteed theory includes the idea of capital service inputs which has come to be interpreted to mean that the firm has a clear-cut decision with regard to capital investment or equipment decisions. Following this type of analysis, Scitovsky[11] defines capital as the current capital funds required to bridge the time lapse between factor input and product output, again a peculiar definition which imposes many conditions on the capital markets. However, the marginal productivity theory is particularly vague concerning the relationships which exist between decisions involving the services of capital and the decisions involving the quantity and timing of capital investment.

The immediate classical post-war work on investment theory divides itself into two areas, and two main

---

[11]T. Scitovsky, Welfare and Competition, (Chicago, Richard D. Irwin, Inc., 1951), pp. 208-209.

individuals, namely George Terborgh and Fredrick and Vera

Lutz.[12] The Terborgh school is concerned with the actual

process of purchasing of equipment and setting up guidelines

for the replacement of physical equipment. The Lutzes, on

the other hand, are more concerned with integrating capital

theory as developed by the neo-classical economist with the

theory of investment for the individual firm. Both theories

and treatments are found somewhat wanting in that the Lutzes

argue that the moment capital goods are introduced, produc-

tion function analysis can no longer be employed unless time

is introduced as an independent variable, but however, in

the analysis of their book, they make little use of time or

the concept of production function.[13] Terborgh, on the

other hand, is concerned only with the replacement of parti-

cular machines and various other items making little to no

---

[12]Concerning equipment analysis and the theory of equipment replacement, the literature has developed around George Terborgh, Dynamic Equipment Policy, (New York, 1949); for the theory of investment of the firm, see Fredrick and Vera Lutz, The Theory of Investment of the Firm, (Princeton, New Jersey, The Princeton University Press, 1951).

[13]Fredrick and Vera Lutz, The Theory of Investment of the Firm, (Princeton, New Jersey, The Princeton University Press, 1951), p. 7.

use of production theory.[14]

In the same vain as the Lutz's attempt to unify
capital and production theory, V. L. Smith in his work
attempted to make explicit the role of technological relations between the theory of investment and the theory of
production. He attempted to develop the empirical and
theoretical interdependence which exists between the short-
run production decision and the long-run investment decision.

In the development of his model, Smith sets up a
single production function of the form $y = f(x_1, X_2, X_3)$,
where y is the flow of current output, x_1 is the flow of
current input, and X_2 and X_3 are the stock of durable goods,
or capital inputs. In his model, the basic production decision for any given level of output involves an optimal
balance between x_1, the flow of current inputs, and X_2 and
X_3, the stocks of capital inputs. Hence, the production
decision of y is inseparable from the investment decision
of how much of X_2 and X_3 to acquire in relation to x_1, the
flow of current input.

---

[14]George Terborgh, Dynamic Equipment Policy, (New
York, 1949); and The Discounted Cash-Flow Method of Invest-
ment Analysis; and Introduction to Business Investment
Analysis, (New York, 1958); Studies in the Analysis of
Business Investment Projects, (New York, 1960); Investment
Policy, (New York, 1964); Administration of Investment Policy;
Control of Capital Equipment; Business Investment Management,
(New York, 1967).

In the first part of the book, Professor Smith is concerned with attempting to relate the traditional production relation of actual production conditions, à la Chenery.[15] In this section he provides many examples of what has come to be called "engineering economy." Therein, he has examples of transmission, batch reactor hemical processes, and other long-run production functions.

From these long-run production functions, he obtains his long-run cost function, and his second theme of his study, the irreversibility of long-run production and cost function. His typical cost function is a declining long-run marginal curve showing increasing returns to the capital goods as the capital goods are varied over the continuum of possible scales. Once a particular scale is selected, the virtual LMC is no longer relevant; with the given durable input, the firm is now contrained to operate on a short-run marginal cost curve, SMC. Assuming a rise in demand, there exists three possibilities for achieving the possible new higher output: (1) for small changes, movement along SMC; (2) for intermediate changes, it is optimal to

---

[15]In V. L. Smith's Investment and Production, (Cambridge, Mass., Harvard University Press, 1961), pp. 3-7, is given the engineering production function for gas transmission originally derived by Hollis Chenery.

install a second facility next to the first; (3) for very large changes, the falling LMC may justify swapping existing facilities, and to invest in a new long-run facility by which optimal production would be attainable.

The investment analysis begins with a single piece of capital that never deteriorates and is always used once purchased. The analysis is gradually generalized by the consideration of capital goods with a fixed finite life and obsolescence, and thus the replacement problem. Further, a number of capital facilities are introduced working in parallel, and the case of fluctuating demand is considered with the durable input having a variable life.

The decision rule which Professor Smith develops is that of the allocating of capital among alternative investments according to the pay-off requirement. He develops the theory of capital allocation in terms of net profit pay-off and the cash flow pay-off concepts. By translating a set of Kuhn-Tucker marginal profitability rules for optimal capital allocation into a set of marginal cash flow pay-off rules for optimal capital allocation, it is shown that the pay-off rules of business practice have the form and structure of optimal rules.

The analysis aside of being difficult and cumbersome in terms of mathematics is questionable. That is,

Smith does not achieve a general useable theory of investment and production and capital replacement. Generally
working within the restrictive assumptions that he outlines,
it is possible to reproduce his analysis. But as for a
general applicable and useable investment-replacement theory
for the individual firm, the book fails.

This is not to say that Smith's work is not
important or a contribution. Without a doubt it is both,
and the sections on production and engineering economics are
without doubt the best exposition in economic literature.
However, what Smith's work exemplifies is the lack of
applicability of the classical theory of the firm to attain
a workable theory of investment production and replacement,
or at most a cumbersome and very narrow theory. Thus, the
economist must look outside the classical theory in order
to obtain workable generalized results concerning the theory
of investment.

Two directions have been taken by economists in
this search. First, there are those who believe that the
new programming techniques will not only be applicable, but
will give the economist the answers he seeks. The second
direction are those who have recently turned to optimum
control theory as a methodology for finding the answers to

capital investment theory.[16] In the ensuing review we will be concerned primarily with the programming techniques since this is the direction of the work which follows and the methodology which has influenced the work of Chapters 6, 7, and 8.

Certainty Models and Programming

Probably the most influential article to follow Dean's work was the well-known Lorie and Savage article. Thus, the influence of this article is our take-off point for the new capital theory.[17] Basically the Lorie-Savage article shows that Dean's rate of return criterion fails whenever: (a) the projects being evaluated are not independent; (b) the sum total of capital expenditures is constrained in more than one period; (c) the stream of returns on a project contains alternations in sign.

---

[16]In Chapter 5 we have discussed Thompson and George's contribution via the calculus of variations method. More recent work is being done by Professor M. El Horidi on an optimal control theory approach to investment problems. Also Professor Smith and Quirk recently published a paper, "Dynamic Economic Models of Fishing," which holds much hope for use in further analysis of the classical investment problem. See J. P. Quirk and V. L. Smith, "Dynamic Economic Models of Fishing," Research Papers In Theoretical and Applied Economics, no. 22, Department of Economics, University of Kansas, (Lawrence, Kansas, June 1969).

[17]J. H. Lorie and L. J. Savage, "Three Problems in Capital Rationing," Journal of Business, (October 1955), Vol. 28, no. 4, pp. 229-239.

The problem as formulated by Lorie and Savage is that a firm is confronted with a variety of possible investment projects with a fixed capital budget. The cash flow associated with each project is assumed to be known and the firm's cost of capital is assumed given and assumed to be independent of the investment decision. Given these assumptions, it is then possible to compute for each investment decision a net present value defined as the algebraic sum of the elements of its stream of each receipts and outlays discounted by the cost of capital. The problem then is to select those projects which lead to the highest present value for the firm.

In general form the problem can be generalized to any number of time periods and can be formulated as a linear programming problem:

209
$$\text{Max} \sum_{j}^{n} b_j x_j$$

210
$$\text{s.t.} \quad c_{tj} x_j \leq C_t$$
$$0 \leq x_j \leq 1$$

where c_{tj} and C_t denote the costs of projects and the budget ceilings in year t, and b_j is the present value of all revenues and costs associated with the individual projects, and x_j represents the fraction of project j undertaken.

Some X_j are then the values for the optimal solution to the problem. The model handles the problems associated with indivisibilities in that the linear program looks at all possible combinations of projects and selects that one for which the present value is a maximum. Also, the upper limit of unity on each x_j assures us that not more than one of any project will be included in the final program. The omission of this would lead to allocating the entire budget to multiples of the best projects. What the model fails to accomplish is to eliminate fractional projects from the solution since that would involve non-linear restrictions requiring the X_j to be zero or one.

Weingartner,[18] set the Lorie-Savage problem in the context of integer programming problems and thus eliminated the problem of multiples of best projects. That is, the solutions which Lorie-Savage obtained from their model do

[18]For a summary and excellent in-depth treatment of the Lorie-Savage article, see H. Martin Weingartner, Mathematical Programming and the Analysis of Capital Budgeting Problems, (Chicago, Markham Press, 1967). This book contains an in-depth analysis of Lorie-Savage and centers on the problems they introduced, examining these in the context of programming models. Also an excellent review of the literature exists in H. Martin Weingartner's "Capital Budgeting of Interrelated Projects: Survey and Synthesis," Management Science, (March 1966), Vol. 19, no. 7, pp. 485-516. As can be clearly seen, my review depends heavily on Weingartner's book and his review article.

yield useful results, but by use of integer programming to deal with the problems of indivisibility of investment projects, more useful results are obtainable.

Given the net present value of a set of independent investment alternatives, and given the required outlays for the project in t time periods, then the problem is to find the set of projects which maximizes the total net present value of the accepted projects while at the same time satisfying the constraints which exist in budget outlays in each of the t time periods. If we let b_j be the net present value of project j when discounting is done by the appropriate rate of interest, and let C_{tj} be the required outlay for the jth project in the tth period, and let C_t be the maximum permissible expenditure in period t, then X_j is defined as the fraction of project j accepted, and X_j is required to be either zero or one. Then the entire model may be written as:

211
$$\text{Max} \sum_{j=1}^{n} b_j X_j$$

subject to:

212
$$\sum_{j=1}^{n} c_{tj} X_j \leqq c_t$$

X_j is either zero or 1.

Given the above formulation of the problem, then the solution depends on finding a good integer programming method to solve the problem. The model as formulated above adds more realism to the problem as originally formulated by Lorie and Savage, in that now the projects j accepted are either accepted in total or are rejected. There are no multiples of best projects. Further, we could put limits on the c_{tj}, the required outlay, that they be non-negative so that the problem has an upper bound and thus has a finite optimum. For the development of an integer program to solve the problem, the reader is referred to Weingartner and the references contained therein.[19]

Kaplan[20] in a recent article integrates Weingartner's reformulation of the problem and incorporates the budget constraint with Lagrange multipliers. Thus, he writes:

213
$$h_i = b_j x_j - \sum_{k=1}^{n} \lambda_k c_{tj} x_j.$$

His method of solution involves selecting arbitrary values

---

[19]Weingartner, Ibid., pp. 161-169, and 487-496.

[20]S. Kaplan, "Solution of the Lorie-Savage and Similar Integer Programming Problems by the Generalized Lagrange Multipliers," Operations Research, (Nov.-Dec. 1966), pp. 1130-1136. This is Kaplan's budget constraint translated into our terminology.

of λ_k, solving the Lagrangian for a maximum and determining
if the budget constraint is met. Each λ_k is then varied and
the solutions that meet the constraints are then examined
for the optimum. As Kaplan points out in his article, solutions may not exist for the optimal point if the solution
is not concave since the multipliers define hyperplanes
tangent to the constraints space, and if the optimum is a
point on a convex surface of the space, there will be no
hyperplane tangent to it. However, Kaplan indicates it is
possible to obtain a near optimal solution much more easily
compared to the difficulties of obtaining a solution by
using an integer programming technique.[21]

Further in Weingartner's[22] presentation, the
Lorie-Savage problem is formulated as a dynamic programming
problem. Basically the difference is that now time sequencing is introduced and a set of sequenced projects are being
considered, the ordering of the projects being arbitrary.
The method consists of determining the list of projects
which would be accepted if the budgets for the t periods
were C_1', C_2' ... C_T', and the selection of projects were

---

[21]Ibid., pp. 1135-1136.

[22]See Weingartner, op. cit., pp. 216-217, 488 and
489.

limited to a finite number of projects i. This is done for
i = 1 ... n, and within each stage i for all feasible
vectors $C' = (C_1', C_2', C_3' ... C_T')$ where feasibility means
that $0 \leq C_t' \leq C_t'$ $t = 1 ... T$. If we define $F_k(C_1 C_2$
$... C_T)$ as the total value associated with an optimal choice
among the first i projects when funds employed are as
defined, then the dynamic programming problem can be stated
as:

214      $F_i(C_1, C_2, ... C_T) = \max$

$$b \, X_i - F_{i-1}(C_1' - C_{1i} X_i)$$
$$(C_2' - C_{2i} X_i) ... (C_T' - C_{T1} X_i)$$

subject to:

215      $C_T' - C_{ti} \not> 0$

and

$$F_o(C') = 0$$

$$x_i = 0, 1$$

Thus, Weingartner extended the Lorie-Savage problem to the multi-period case, and to the multi-state analysis, showing that the model can be solved by linear programming methods and by integer programming methods by including
the integer requirement for the X_j's and by dynamic

programming techniques given a multi-stage series of budgets and projects. Weingartner further shows how the solution of the dual by linear programming can give insights into the cost of the budget constraints. He points out that as a constraint, the rationing of any input (in this case capital) plays a very important part in the analysis. In summary then, the model Weingartner considers is concerned with multi-time periods. However, the capital budget ceiling is known at all times and the start of any investment project is known at all times and the start of any investment project is limited to the present period, or some given fixed future period.

In a similar analysis, an article by Charnes, Cooper and Miller[23] is concerned with the allocation of funds within an enterprise. Basically, the authors set up a warehousing problem of the form:

216

$$\text{Max } \pi \sum_{j=1}^{n} p_j y_j - \sum_{j=1}^{n} c_j x_j$$

[23]A. Charnes, W. W. Cooper and M. H. Miller, "Application of Linear Programming to Financial Budgeting and Cost of Funds," Journal of Business, (Feb. 1959), Vol.32, pp. 20-46.

subject to:

217

$$\sum_{j=1}^{n} X_j - \sum_{j=1}^{n} y_j \leq B - A$$

and

$$\sum_{j=1}^{n-1} X_j + \sum_{j=1}^{n} y_j \leq A$$

$$X_j y_j \geq 0 \text{ for all } j$$

where they define:

B = fixed warehouse capacity;

A = initial stock of inventory in warehouse;

X_j = amount to be purchased in period j;

y_j = amount to be sold in period j;

p_j = price prevailing in period j;

C_j = purchase price per unit prevailing in period j.

The objective function is that of maximizing profits subject to the constraints that no more can be bought than can be stored in the warehouse in any period, and secondly, the amount sold cannot exceed the amount on hand in that period of time, and thirdly, the amount purchased and sold are assumed positive in any time period. Later in the article, the authors introduce financial constraints into the model, that is, they allow for funds available for purchase of inventory and also allow for borrowing and lending

to purchase inventory. Further, they show how the solution
of the dual and the economic interpretation of the dual
variables have influenced not only an additional warehouse
capacity but also on the cost of capital.[24] Note that in
the objective function, there is no discounting of future
profit and that the constraint or borrowing is fixed.

Research has also been conducted on the physical
aspects of the investment problem, that is, works concerned
with plant location, plant size, and timing of investment.
For example concerning the plant size problem, Chilton[25]
made a study of different types of chemical plants and
found what has come to be known as the six-tenths rule:
that is,

218 $$\text{investment cost} = k(\text{size})^{.6}$$

Allan S. Manne[26] recently undertook a study of
plant size and investment timing and found that Chilton's
equation 218 was a very good approximation of investment
cost. Manne continues to discuss in this work a more

----

[24]Ibid., p. 25.

[25]Chilton, C. H., "The Sixteenth Factor Applied to
Complete Plant Cost," Chemical Engineering, (April 1950),
pp. 112-114.

[26]A. S. Manne, Investment For Capacity Expansion,
Size Location and Time-Phasing, Cambridge, Mass., The Harvard
University Press, 1967).

general problem, that is, the problem of plant location
covering several geographical areas. He concludes that it
is advantageous to build large plants successively in different areas and to meet the different demands by trans-
shipment of the goods from one region to another.

In an earlier model, Manne[27] uses plant size as a
decision variable to determine the optimum building interval
to meet an expanding demand. In the absence of any capital
constraints, the excess capacity of plants can be treated
as an inventory item and the methodology which can be
employed is that of the formulation of an economic lot size
problem with and without backlogging. Manne uses as his
objective function:

219        $\text{Min } C(x) = kx\alpha + e^{-rx}$

where:

kx = installation cost relation;

r = discount rate.

His objective is to minimize the cost of investment and the
opportunity cost on idle capacity. Manne assumes a linear
growth in demand and concludes that the cost in relatively
insensitive to the building interval. His model applies

[27]A. S. Manne, "Capacity Expansion and Probabilistic Growth," Econometrica, Oct. 1961), Vol. 29, pp. 632-
649.

particularly well to conditions in the gas and oil pipelines industry, the telephone industry, highway construction and electric power generators.

I. McDowell considers a similar type problem in his 1960 article, but instead of being concerned with linear growth in demand, he assumed demand grows exponentially.[28] Again McDowell is concerned with the utilities industry electricity, telephone, highways, pipelines, etc. Like Manne he formulated his model to minimize cost of investment and the opportunity cost of idle capacity. The solution which he offers is somewhat clumsy in that he relies on direct emmeration and varying parameters to solve the problem.

A restatement of McDowell's paper was made by D. J. White in the same Journal, however, casting the problem in a dynamic programming light.[29] White sets the problem as follows: first, the planning horizon is known and is N. F(s) is defined as the cost of providing for service demand for the time periods s to N, beginning at installment

[28]I. McDowell, "The Economical Planning Period for Engineering Works," Operations Research, (1960), Vol. 8, pp. 533-42.

[29]D. J. White, "Comments On a Paper by McDowell," Operations Research, (1961), Vol. 9, pp. 580-584.

time s and using an optimal policy over the remaining years
and discounting the returns. Then the problem is:

220 $$F(s) = \min_{s \leq r \leq n} \left[b + a \left\{ e^{kr} - e^{ks} \right\} + e^{-c(r-s)} \right]$$

where the interest rate, c, has been introduced as a
discount factor instead of an interest factor and thus per-
mitting discounting of all costs back to the initial period.
It is now possible to compute $F(N)$, $F(N-1)$, $F(N-2)$ ... $F(0)$,
or vice versa. The remainder of the paper is given to an
investigation of the characteristics of the optimal policy.[30]

The next study of importance in the analysis of
capital budgeting theory is that of Baumol and Quandt in
1965.[31] Basically, the model Baumol and Quandt developed is
to introduce the cost of capital into the capital budgeting
analysis as well as into capital investment theory. They
are in effect attempting to unite the so-called classical
theory of capital to the mathematical techniques of capital
budgeting.

The development of the model by Baumol and Quandt

---

[30]Ibid., see the paper for detailed results of the
Lemmas proved.

[31]W. J. Baumol and R. Quandt, "Investment and Discount Rates Under Capital Rationing--A Programming Approach,"
Economic Journal, (June 1965), Vol. 75, no. 298, pp. 317-329.

is heavily dependent on the works of Weingartner (1963) and Charnes, Cooper, and Miller (1959). They assume that funds borrowed for investment purposes can be borrowed, and that there can be future starting dates for some of the investment alternatives. They assume further that the capital funds available are known, fixed, and given from outside the model. This essentially means that the investment alternatives considered are independent of the remaining organization. Baumol and Quandt develop a number of models initially in the paper as an introduction to their analysis until finally they arrive at the model they want to consider. The last model considered has an objective function with the utility of cash throw-offs included.[32] The final model is given as:

221 $$\text{Max} \quad \sum U_t W_t$$

subject to:

222 $$\sum a_{jt} X_j + W_t \leq M_t$$

$$X_j \geq 0$$

$$W_j \geq 0$$

where they had defined previously the notation used as:

---

[32]Ibid., p. 326.

U_t = fixed utility of a dollar in period t;

W_t = amount withdrawn -- cash throw-off;

a_{jt} = investment required in j^{th} alternative;

X_j = proportion of the jth alternative accepted;

M_t = budget limitation in the tth period.

They immediately formulate the dual of the problem as:

223
$$\min \sum_t p_t M_t$$

subject to:

224
$$-\sum_t a_{jt} p_t \geq 0$$

$$p_t \leq U_t$$

where again p_t is the discount rate. In solving the dual,
they state that the ratios of utility for any two periods
t and t', must be less than or equal to the ratios of the
dual variables p_t and p_t' or as they show:

225
$$p_t = U_t$$

$$p_t' = U_t'$$

Then:

226
$$\frac{U_t}{U_t'} = \frac{p_t}{p_t'}$$

and this ratio is interpreted by them as the production
opportunity discount rate, and as such represents the "cost

of capital," or the disutility of deferring withdrawal of funds.

In none of the models thus far considered have any of the authors been concerned with the interdependence or possible interdependence of investment alternatives. Also, each model has explicitly or implicitly assumed a perfect market for investment funds. It is thus necessary before considering models of uncertainty to consider the implications of these assumptions and the consequences of relaxing them.

When projects are related to each other, these relationships must be taken into account. Weingartner has distinguished four forms of relationships which can exist. He defines:

1. Independent projects: when the worth of individual investment proposals is not profoundly affected by the acceptance of others.

2. Mutually exclusive projects: when acceptance of one proposal in such a set renders all others in the same set unacceptable or even unthinkable.

3. Contingent projects: when acceptance of

33Ibid., pp. 326-327.

one proposal is dependent on acceptance
of one or more other proposals.

4. Compound projects: when contingent projects
   are combined with the projects on which
   they depend so that the independent project
   and the compound project may be treated as
   mutually exclusive alternatives.[34]

The first two of these relationships have been handled by
Lorie-Savage and Weingartner in their respective works.

To make the extension from
necessary in the linear programming case merely to add one
equation for each set of mutually exclusive alternatives,
namely:

227
$$\sum_{j \in J} x_j \leq 1$$

where the summation is over the indices corresponding to the
set j of mutually exclusive alternatives.

The situation of contingent alternatives can be
handled by an extension of the methodology of the mutually
exclusive method. This is possible by constructing compound
projects which include contingent alternatives. A way of

[34]H. Martin Weingartner, _Mathematical Programming_
and the Analysis of Capital Budgeting Problems, (Chicago,
Markham Press, 1967), pp. 10-11.

stating a contingency restriction would be as follows: If

project m is desirable if and only if project k is adopted,

but not otherwise, then:

228 $\qquad X_m \leq X_k$

If $X_k^* = 0$, then of necessity $X_m^* = 0$. The usual requirement

is that X_k 1 and this implies likewise that X_m is $\leq$ 1

because $X_m \leq X_k$.[35]

The simplest way in which the assumption of per-

fect capital markets may be dropped is to assume that their

is an upper absolute limit on the amount of debt outstanding

for a firm at any given moment in time. Charnes, Cooper and

Miller impose a liquidity constraint on the financial manage-

ment of the firm in the form of a minimum cash balance.

There are any number of possible ways to impose limits on

borrowing.[36]

The model may be set up as a basic horizon model,

as done by Weingartner,[37] with the constraints set now

---

[35]The effects of contingency relations are pursued
by Weingartner in section 8.3 and 8.4 and the interested
reader is referred to those sections.

[36]See N. H. Jacoby and J. F. Weston, "Factors
Influencing Managerial Decisions in Determining Form of Bus-
iness Financing," Conference on Research and Business Finance,
(New York, 1952) (NBER), pp. 145-146.

[37]See Weingartner (1963) Chapter 8 for basic
horizon model.

including the borrowing restrictions. [38] Weingartner in his
analysis of imperfect capital markets, demonstrates that the
common criteria for investment decisions are not appropriate
tools for choosing among investments when there are limits
imposed on borrowing at a given interest rate. In the analysis he develops tools necessary to consider this new uncertainty, that is, the rising supply curve for funds, absolute
borrowing limits, nonrenewable short-term bonds, and equity
financing and places these effects in the linear and integer
programming framework to obtain the optimal investment
policy. This analysis will be used somewhat in Chapter 8
of this thesis, and thus for continued detailed analysis of
the methodology used by Weingartner, the reader is referred
to his 1963 book or to Chapter 8.

Uncertainty Models

To the present we have directed our attention to
studying the literature surrounding capital investment in
deterministic models. However, much recent work has been
framed in models containing risk and uncertainty. Often we
use the term risk and uncertainty in the same breath. First
then it is important to understand what is meant by risk and

---

[38]For an analysis of the models developed with
imperfect capital market, see Weingartner, Chapter 9.

264

uncertainty. First, uncertainty refers to a situation in
which future outcomes are imperfectly known. A situation is
said to involve risk only when the probabilities of possible
outcomes are known. To qualify as a risk situation then,
an experiment must be repetitive in nature, possess a fre-
quency distribution from which observations can be made and
about which inferences can be drawn by objective, statistical procedure. Uncertainty, on the other hand, is said to
be present when the experiment in question cannot be carefully replicated by other persons at other times and places.
That is, the situation can be said to be unique. Its fre-
quency distribution then cannot be objectively specified.

Here, where the concern is with deterministic
model, only a brief survey of the literature on capital
investment, capital budgeting models incorporating risk or
uncertainty will be analyzed.

In an influential article published in 1952, Harry
Markowitz demonstrated that a preference for expected value
and an aversion toward risk can be shown to lead to the
development of a diversified investment portfolio.[39]

---

[39]See Harry Markowitz, "Portfolio Selection,"
Journal of Finance, (March, 1952), p. 89, and also in H.
Markowitz', Portfolio Selection: Efficient Diversification
of Investments Cowles Commission, Vol. 16, (New York, John
Wiley and Son, Inc., 1959).

Briefly, the Markowitz model can be formulated as below:
first, let π_1 represent a price index for the ith security,
having unit value during the current time period. Assume n
such securities exist. Then

229
$$U_1 = E\ (\pi_1) \qquad i = 1 \ldots n$$

defines the ith security's expected price index at the end
of the planning horizon in question. Also

230
$$\sigma_{ij} = E\ (\pi_1 - U_1)(\pi_j - U_j)$$

can be defined as the anticipated covariance of securities
i and j about their expected values, with anticipations
being based on past experience with the technique by which
the set of expectations U_1, $U_2 \ldots U_n$ are generated. Define
$V_1 \geqq 0$ as the total proportion of assets which an investor
puts into security i. V_1 is such that $\sum_{1}^{n} V_1 = 1$ and

231
$$N = \sum_{i=1}^{n} V_i \pi_i$$

is the portfolio itself.

The application of expected value operators to the
weighted sum shows that a portfolio's expected value is
nothing but the same weighted sum of each security's individual expected value, that is:

232
$$E(N) = \sum_{i=1}^{n} V_i \pi_i \qquad \text{for all i} \quad 1 \ldots n$$

and that its variance can be expressed:

233 $\qquad V(N) = \sum_{i=1}^{n} \sum_{i=1}^{n} V_i \, \sigma_{ij} \, V_j \qquad$ for all ij.

Decomposing the portfolio's variance into a linear sum of the variances and covariances of individual securities:

234 $\qquad V(N) = \sum_{i=1}^{n} V_i^2 \, \sigma_i^2 - 2 \sum_{\sigma < i} \sum_i V_i \, \sigma_{ij} V_j$

it can be seen that the possibility of a portfolio whose variance is smaller than the smallest variance possessed by a single security depends only on the existence of sufficiently small covariance elements. Therefore, there is no reason why a diversified portfolio cannot yield a variance which is smaller, for almost any given expected value, then that for a single type of asset.

Markowitz also redirects the manner in which an investor is assumed to evaluate his potential investment undertaking. The investor is now expected to consider the effects on the entire portfolio of a change in asset composition. Portfolios then, not individual assets, are the objects which form the set of available investment opportunities.

Markowitz's analysis stimulated research in portfolio and investment analysis, particularly his 1959 work.[40] In 1956 in Econometrica, R. Freund developed an

interesting variational model.[41] Freund specifies his

utility money function as based on the Markowitz analysis

235                    $U(X) = 1 - e^{X\ A}$

where he defined:

    A = investor's subjective coefficient of
        risk aversion;

    X = the investments expected return as a
        stockhastic variant

Freund further assumes the investment's return is normally

distributed about its mean. Using these assumptions, and

going on to postulate expected utility maximization as the

investor's decision criterion, he forms his objective

function to be maximized as:

236            $E\left[U(X)\right] = \int_{-\infty}^{\infty} (1 - e^{-AX})\ e^{-\frac{1}{2}(x\ \frac{u}{\sigma})^2}\ dx$

Since he has a definite empirical application in mind, he

next sets out to choose an appropriate value for the coeffi-

cient of risk aversion. However, he has no objective criter-

ion to rely on to make this important decision so he makes a

---

[40]Ibid., various sections of the work.

[41]R. Freund, "The Introduction of Risk into a
Programming Model," Econometrica, (1956), p. 258.

stab in the dark for a trial value, and then proceeds to generate optimal portfolios.[42]

His work is important even though he possesses no objective criteria for selecting one risk aversion coefficient over another, and even though his problem is very limited in scope.

Henry Latane[43] was concerned with this same problem in his 1959 paper, that is, the problem of making rational choices among strategies in situations involving uncertainty. Basically he breaks the process of rational decision-making into three steps: (1) deciding upon an objective and criteria for choosing among strategies; (2) filling out a payment matrix; and, (3) choosing among available strategies on the basis of (2) and (1).[44] The paper is concerned with developing an objective and criteria for selecting a risk aversion coefficient which he feels

---

[42]Ibid., p. 258 for the results he obtains for farm crops on an Eastern North Carolina farm.

[43]Henry Allen Latarie, "Criteria for Choice Among Risky Ventures," Journal of Political Economy, (1959), Vol. 67, pp. 114-155.

[44]Ibid., p. 145.

Markowitz and Freund and Savage[45] failed to do. The criteria for choice between risk and safety in portfolio can be summarized as that which the wealth-holder adapts as the maximum change subgoal, that is choicing the portfolio that has a greater probability P of being as valuable or more valuable than any other significantly different portfolio at the end of n years. It is shown that the portfolio having a probability since G depends both on the mathematical expectation and the variance of the probability distributions of returns. When G is maximized, there is no change of ruin if the probability beliefs are correct. Thus G falls in the realm of accepted reactional behavior.[46]

Utilizing the work undertaken in portfolio analysis, D. E. Farrar[47] undertook to study uncertainty in decision-making models of investment. Basically he relies on two assumptions: (1) an investor's utility of money function is positively sloped and concave downward;

---

[45]See M. Friedman and L. Savage, "The Utility Analysis of Choice Involving Risk," Journal of Political Economy, (1957), Vol. 56, pp. 279-304 for a development of their model.

[46]op. cit., pp. 146-147 and 154-155.

[47]D. E. Farrar, The Investment Decision Under Uncertainty, (Englewood Cliffs, N. J., Prentice Hall, 1962).

(2) his investment strategy is the maximization of expected
utility. Using these assumptions, the investor's utility of
money schedule can be expanded by Taylor's series about its
mean.[48]

237 $$U(X) = (U_n) + U'(U)(X - u) + \frac{U'(U)}{2} (X - U)^2 + \ldots$$

where X is the investor's wealth or net worth as a stochastic variate. Subjective probability beliefs are held for
the likelihood that it will assume alternative possible
values. The expected value of this variate is:

238 $$U = E(X)$$

and 239 $$U'(U) < 0$$

expressed the diminishing marginal utility of wealth assumption which he has implicitly assumed.

Applying expected value operators to each side of
Equation 237, the investor's expected utility may be
expressed:

240 $$E\left[U(X)\right] = U(U) + U'(U)E(X - M) + \frac{U'(U)}{2}$$

$$E(X - U)^2 + \ldots$$

However, it is apparent that the expected value of $(X - U)^2$
is the variance of X, and as $E(X - U) = 0$, the middle terms
vanish so the series can be rewritten:

------

[48] Ibid., pp. 19-22.

241 $$E\left[U(X)\right] = U(U) + \frac{U'(U)}{2}\sigma_x^2$$

This then is the objective function which the rational investor attempts to maximize.

Farrar then attempts an empirical justification of the model using mutual fund data. Consequently, he finds his model is a good predictor of optimal portfolio policies of mutual funds.

The contribution of the analysis is the extension of orthodox utility theory into the uncertain world; and secondly, the development away from the single asset domain to one of multi-dimensional mathematical programming. Likewise, by attempting to evaluate an investors subjective attitude toward risk, the tool of mathematical programming is transformed into a decision unit model which can be empirically tested.

In a more recent article by Cord,[49] the capital investment problem considers capital rationing and independent alternatives. Cord uses the internal rate of return methods to determine the per unit return, P_i, in his terminology. With each return, he associates some variance as a

[49]J. Cord, "A Method for Allocating Funds to Investment Projects When Returns Are Subject to Uncertainty," Management Science, (1964), Vol. 10, pp. 335-341.

measure of risk V_i. Also, he assumes that the decision maker
has some absolute level of risk which he is not willing to
exceed. The model is formulated as follows:

242 $\qquad$ Max $\qquad T = \sum_i^n p_i X_i$

subject to:

243 $\qquad \sum_i^n I_i X_i \leq I$

$$\sum_i^n w_i X_i \leq V$$

$$X_i = o, 1$$

where he defines:

$\qquad P_i$ = per unit return on alternative x;

$\qquad X_i$ = proportion of the xth alternatives selected;

$\qquad I_i$ = investment in the ith alternative required;

$\qquad I$ = Funds available for allocation;

$\qquad V$ = subjective limit on variance;

$\qquad V_i$ - variance of the expected return on the ith project.

Cord immediately converts the model to one using Lagrangian
multipliers:

244 $\qquad$ Max $\sum_1^n P_i X_i - \sum_1^n w_i X_i$

subject to the same constraints. In this form he is now able to treat the problem as a multi-stage allocation problem in the same form of the knapsack problem and is able to use dynamic programming techniques.

Then for a given investment level and various budget levels I' as the state variables, he expresses the stages process as follows:

245

$$F_1(I') = \max_{0 \le X_1 \le 1} (p_1 X_1 - \lambda w_1 X_1)$$

$$F_2(I') = \max_{\substack{0 \le X_2 \le 1 \\ 0 \le I' \le 1}} p_2 X_2 - \lambda w_2 X_2 + F_1(I' - I_2 X_2)$$

. . . .

$$F_n(I') = \max_{\substack{0 \le X_n \le 1 \\ 0 \le I' \le 1}} p_n X_n - \lambda w_n X_n + F_{n-1}(I' - I_n X_n)$$

If $\lambda = 0$, then the variance constraint is eliminated, that is, the variance is allowed to be infinite. As λ increases, the bound is corresponding lowered. Thus, Cord's method of solution solves the problem for successive values of λ until the constraint $\sum_1^n I_i X_i \le I$ is met.

274

A recent article by Hartung[50] develops further this model expressed by Cord in his 1964 paper. Again the decision variables are investments to be made in a single period under capital rationing. The alternatives are independent, and the objective function is to maximize the present value of cash flows. With each investment, he associates a risk or variance and an upper limit on the total variance. Hartung differs from Cord in that he uses several costs of capital the firm may have and orders these costs with an associated range of funds available. These cost of funds are entered into the objective function as a reduction on returns, and dynamic programming is used as the solution technique.

Another major work on risk in capital investment was that undertaken by Hillier. Hillier points out that risk, if treated at all, tends to be treated in several simplifying ways that may not always be valid. First, is to assume that for each random variable, its expected value will prevail. Secondly, risk has been introduced by adjusting the interest rate either for the cutoff level in the internal rate of return method, or directly in evaluating net present value.

[50]P. H. Hartung, "Capital Budgeting With Several Costs of Capital," 13th T.I.M.S. Conference, September 1966.

A third method is to perform a sensitivity analysis involving revision of uncertain estimates of prospective cash flows and observing the sensitivity of the measure of merit. Hillier then suggests that expected utility would be an ideal measure of merit from a theoretical point of view, and has, in fact, been selected by some of the works reviewed herein. However, he suggests that this measure suffers from the operational difficulty of being extremely difficult to measure and highly variable over time.[51]

Hillier then suggests that a good compromise between the above three methods is to use the expected value and standard deviation of the rate of return as decision parameters. He assumes that net cash flows are normally distributed. He then performs his own sensitivity analysis in that the cash flows can vary from complete independence to complete correlation, reality being somewhere in between. Further he argues that production costs are apt to be independent, and that thus any variation is due to random disturbances. However, sales and revenues are likely to be highly correlated from one time to another. He then develops

[51]F. S. Hillier, "The Derivation of Probabilistic Information for the Evaluation of Risky Investments," Management Science, (April 1963), pp. 443-457.

his models in terms of present worth of an investment in terms of independence, partial correlation, and complete correlation. The result is that the total variance is a minimum if the returns are independent and maximum for complete correlation.[52]

Van Horne[53] in a recent article, uses the tools formulated by Hillier combined with the methods of Markowitz and subjective conditional probabilities and makes application with these tools to the security-portfolio problem.

Another major work that needs to be considered is the works of Naslund.[54] The first paper (1962) presents a multi-period model of rational investment in the stock market, and decision rules are formulated such that they generate the highest expected return on investment while maintaining the risk constraints. In his second paper (1966) he is directing his model towards a method of chance

---

[52]Ibid., p. 446.

[53]J. Van Horne, "Capital Budgeting Decisions, Involving Combinations of Risky Investments," Management Science, (Oct. 1966), pp. B84-B92.

[54]B. Naslund and A. Whinston, "A Model of Multi-Period Investment Under Uncertainty," Management Science, (Jan. 1962), pp. 184-200. B. Naslund, "A Model of Capital Budgeting Under Risk," Journal of Business, (April 1966), Vol. 39, no. 2, pp. 257-271.

constrainted programming solution. His investment alternatives are classified as independent, and to the decision
variables he adds the possibility of lending and borrowing
in each period. Further, a limit is made in the funds
available.

In his discussion of risk, Naslund points to the
desirability of using utility theory as many before him have,
but also recognizes the operational difficulties resulting
from its use. He mentions the certainty equivalence requirement approach and under what conditions it should be used.
To use chance constrainted programming requires that the
decision makers know and be willing to specify the probability level at which the constraint must be met. Thus,
each constraint is of the form:

246 $$P(a \leq B) \geq \alpha$$

that is, that the probability that α is less than or equal
to B will always be greater or equal to some level α.

The problem can then be stated as the maximization
of the expected value of the firm at horizon T while requiring that the money available for investment is the money
generated by investment projects in previous periods, plus
the interest rate earned in previous lending, minus the
interest paid on previous borrowing, plus the amount borrowed,

minus the amount lent, plus the amount generated by the rest of the business. It is possible to state this in simplier form mathematically as:

247     $\text{Max} \quad E \left(\sum_i \hat{a}_1 X_1 + N_T - W_1 \right)$

subject to:

248     $P_r \left(\sum_i a_{1j} X_j + V_1 - w_1 \leqq D_1 \right) \geqq \alpha_1$

$$P_r \left(\sum_{t=1}^{t} a_{1j} X_j - \sum_{t=1}^{t=1} V_{1r} + \sum_{t=1}^{t=1} w_{1r} + V_1 - w_1 < \right.$$

$$\left. \sum_{t=1}^{t} D_1 \right) \geqq \alpha_1$$

$X = 0, 1$

$w_T, V_T \geqq 0$

where he has defined:

$a_1 =$ horizon value of all flows for investment i;

$V_T =$ the amount lent in period t at interest rate V;

$W_t =$ the amount borrowed in period t at interest rate r;

$a_{1t} =$ flow of money of the ith alternative in period t;

$X_j =$ proportion of the jth alternative selected;

$D_t =$ amount of funds generated outside the set of alternatives in period t.

To develop his deterministic equivalent of his

model, Naslund invests the probability constraints and
replaced the anticipated flows by their means and variances.
He likewise assumes they are normally distributed random
variables. His general constraint set then becomes:

249

$$\sum_i U_{1j}X_j + V_1 - W_1 + (\sum_j \sigma_{1j}^2 X_j^2) \; F^{-1} \; \alpha_1 \le D_1$$

$$\sum_{i=1}^{t} {}_1 U_{1j}X_j - \sum_{i=1}^{t-1} V_{1j} + \sum_{i=1}^{t-1} W_{1j} + V_t - W_t +$$

$$(\sum_j \sigma_{tj}^2 X_j^2)$$

$$F^{-1} \alpha_t \le \sum_{t=1}^{t} D_1$$

The first five terms of the latter expression Equation
gives the mean net flow of funds. The second part represents a transformation of risk aversion into safety margin
above what is required on the average to meet the budget
constraints.

Naslund in solving the dual by using the Kuhn-
Tucker conditions, develops acceptance criteria for the
investment. Further, he prints out his model in detail,
which is non-linear in its constraints, and suggests some
methods which could be employed in obtaining solutions. At
the very end of the paper, he does perform some sensitivity
analysis to show the effect of variance, the rate of return,
and the probability level on the solutions.

Summary

In this review of the current literature on capital investment problems, most of the considerations which will be incorporated in the problem formulated in Chapter 8 have been included. The models we have reviewed all assume that a fixed known amount of funds was available for every time period. Likewise, most of the models reviewed assumed that the investment alternatives were independent of one another and the existing structure.

In the model of Chapter 8, following Naslund, we would like to consider problem of time and sequencing of investment projects, and likewise sequence and time these interrelated projects in terms of a dynamic budget constraint wherein the funds available are not always known or fixed. The method of solution, as we shall see, is a branch bound algorithm which looks at the total enumeration of interrelated projects.

CHAPTER 8

LINEAR AND DYNAMIC FORMULATIONS

OF THE INVESTMENT PROBLEM

Introduction

In the previous Chapters 5 and 6, the rates of
production for the reservoir were the unknowns to be solved
for and once determined, assuming that the cumulative recovery of the reservoir was known, it was possible to arrive at
the well spacing and the optimal investment pattern via these
optimal rates of production. Essentially then we were concerned with finding the production decline curve for the
reservoir given the physical characteristics, and this production decline curve was defined to be some optimal average
rate of production. Thus, the linear models formulated were
designed to yield this optimum rate of production, and the
dynamic model gave us the optimum rates of production in a
series of stages, assuming that each prior stage was maximized. We were able then through dynamic programming to
incorporate the interdependencies of the physical and the

economic aspects of the problem to determine the optimum solution to the problem. The physical aspects through the introduction of the decision variable pressure, and the economic aspects through the introduction of the economic decision variable of profit and investment, were analyzed through stages to give us an optimal well operation rate along with an optimal well spacing program.

The models we would now like to consider differ from those studied previously. First, it is assumed that the production from a given well will follow a set but unknown production decline curve. The well continues to follow this decline curve until abandonment or until the time that it is shut-in. The results of this assumption are that we now no longer need to be directly concerned with the pressure constraint, or with any of the other physical characteristics. We need not introduce an influence function between wells since the production decline curve for the wells and the reservoir will follow the set decline curve.

The jth well will then follow the decline curve given by Equation 250.

250
$$q_{scj} = d_j.$$

This is simply a decline curve. It is suppose to represent the production decline that actually occurs in practice in

a particular well on a given reservoir. It is assumed that
a particular well will begin following Equation 250 as soon
as it is completed, and that it will continue to follow this
decline curve until it is finally shut-in. Hence, the flow
characteristics of all j wells in the reservoir can be completely specified by the equation:

251
$$\sum_{j=1}^{N} \sum_{t=1}^{T} q_{scj}(t) = d_j(t)$$

The Model Assumptions[1]

There are a number of assumptions that are necessary for the models which we will develop, both linear and
dynamic. First, the cumulative reserves for the reservoir
are known, that is the size, the dimensions, the net pay
thickness and the other physical characteristics of the
reservoir are given. Also, there exists a distinct number
of geological and geographical production points, and there
exists some method for allocating the facilities and the
producing wells. Likewise, we assume that the production
decline curve for the reservoir is given by Equation 251.
All wells on the reservoir are taken to be identical in all

---

[1]For a linear model of a similar nature, the
reader is referred to: J. S. Aronofsky and A. C. Williams,
"The Use of Linear Programming and Mathematical Models in
Underground Oil Production," Management Science, (1962),
Vol. 9, pp. 396-406.

their characteristics and the product produced is taken to be homogeneous, that is, only gas or oil is produced, not both. Thus, the problem of joint production and costs is avoided.

Further, we now assume that wells have to be drilled, and that in order to drill wells it is necessary to have rig and drilling equipment, as well as other completion materials. First, rigs are necessary for drilling wells, and they may be either purchased or rented on a given reservoir over the relevant time period. We can define for the tth time period that the number of rigs used in that time period must not be greater than the total number of rigs purchased or rented for a specific reservoir in that time period:

$$252 \qquad \sum_{k=1}^{K} \sum_{t=1}^{T} p_k(t) \leqq \sum_{k=1}^{K} \sum_{t=1}^{T} r_k(t)$$

where:

p_k = the total number of rigs in operation in the tth time period;

r_k = the number of rigs purchased or rented during the tth time period.

Secondly, the number of wells drilled during the tth time period must not be greater than the number of wells that could have been drilled based on the number of rigs that operated during the tth time period, whether the wells being drilled were finished or not. Thus, we have the constraint:

253 $$\sum_{j=1}^{n} n_j(t) \leqq p_k \bar{r}_k (t-t-1)$$

where:

$n_j(t) =$ the number of wells drilled at the end of the tth time period;

$\bar{r}_k =$ the number of wells drilled by a rig in a unit of time lapse within the t   time period.

In compliance with the assumptions made, the cumulative number of wells drilled by the end of the tth period cannot be greater than the maximum number of wells that all operating rigs could have drilled considering that one rig can drill only a limited number of wells during any time period, and a limited number of wells during the rig lifetime. Thus, we add the constraint:

254 $$\sum_{j=1}^{N} n_j(t) \leqq a \sum_{t=1}^{T} p(t)$$

where a is equal to the number of wells one rig can drill during the time period t.

Finally, there exists on any reservoir a distinct number of geological and geographical locations for wells to be drilled, and this cumulative number of wells drilled must not be greater than the total number of wells that have a possibility of being drilled during that time period:

255
$$\sum_{j=1}^{n} n_j(t) \leqq N$$

where N is the total number of wells that have geologic as well as geographic possibility on any given reservoir.

Having determined the number of wells that should have been drilled on any given reservoir, then it is possible to determine an instantaneous flow rate for the reservoir. If there are n_j number of wells drilled on the reservoir during the time period t, and if each of the n_j wells follow the production decline curve of Equation 250, then during any time period t the instantaneous flow rate for the reservoir can be given by:

256
$$\phi(t) = \sum_{j=1}^{N} \sum_{t=1}^{T} n_j(t) \, d(t-t-1)$$

If $t = 0$, then $d = 0$ and $\phi(0) = 0$. For any other time period the instantaneous flow rate cannot be greater than the pipeline capacity. This constraint can be expressed as:

257
$$\sum_{j=1}^{N} \sum_{t=1}^{T} n_j(t) \, d(t-t-1) \leqq W_c$$

The cumulative flow at the end of the tth time period can be expressed as:

258 $\qquad R(t) = \displaystyle\sum_{j=1}^{N} \sum_{t=1}^{T} \left[n_j(t)\, d(t-t-1) \quad \Delta t \right]$

From the overall materials balance equation, we know that the cumulative flow cannot be greater than the total amount of oil in place for any given well so that the reserves constraint can be given as in Equation 259.

259 $\qquad R(t) \leqslant V_o \displaystyle\sum_{j=1}^{N} n_j$

where V_o is the initial oil in place that has possibility of being recovered, and n_j are the number of wells that have been drilled.

Recapitulating, Equations 254, 255, 256, and yield us the drilling restrictions imposed on the reservoir in determining the number of wells to be drilled, while 257, 258, and 259 provide us with the flow characteristics of the reservoir given the number of wells that have to be drilled. All the constraints are linear, and thus, we have the traditional linear programming problem, and since only one reservoir is being considered with the possibility of many wells being drilled on it, in the final analysis we have 6T.

The form of the objective function is very much open. It would be possible to formulate it in terms of instantaneous cash flow, discounted cash flow, net present worth, percentage gain on investment, or in some other net earnings concept. Any of the above would fulfill the purpose of maximizing some profit function, however defined.

The instantaneous cash flow to the firm at any point in time is simply defined as the revenues derived from the sale of the product to the pipeline (the consumer) less the operating expenses incurred at the reservoir in obtaining the product, plus the capital investment required to initiate and to develop the reservoir by the individual firm. Thus, the instantaneous cash flow at any time period t can be given by:

260
$$CF(t) = R(t) - C(t) - O(t) \left[\sum_{j=1}^{N} \sum_{t=1}^{T} \left[n_j(t) \, d(t-t-1) \right] \Delta t - \left[d \, r_t - m(p_t \Delta t) - x \, n_j \right] \right]$$

where we have defined:

  $R(t)$ = the revenues received from the sale of the product in time period t;

  $C(t)$ = the operating costs to lift the oil and to transport it to the pipeline terminal;

  $O(t)$ = the production tax as a percent of revenues, and royalties as a percent of revenues, $5R(t) - 1/8 \, R(t)$;

  d = the purchase or rental price of a drilling rig;

$r(t)$ = the rigs purchased or rented during the tth time period;

m = the expense incurred in operating and maintaining the rigs;

$p(t)$ = the number of rigs in operation in time period t;

n_j = the number of wells drilled and in operation in time period t;

x = the completion cost per well.

To arrive a cumulative cash flow over the life of the reservoir, we take all revenues and all expenses over all time periods that the reservoir is in operation and arrive at the concept of cumulative cash flow:

261
$$CCF(t) = \sum_{t=1}^{T} CF(t)$$

If we desired, we could include as a cost to the firm the taxes paid by the firm, and likewise include as a source of revenue the allowances made for depreciation, and in the case of oil and gas, depletion. However, we shall at the present leave out these last two considerations at the present time and introduce them into the dynamic considerations of the model.

In order to consider and to determine an optimal drilling plan for a particular reservoir, the use of cumulative or instantaneous cash flow is not an appropriate yardstick for measuring the net worth of an investment project. Time, the time flow of revenues and costs, is an extremely important element of the analysis which up to now has been omitted. Thus, the cash flow concept used should be discounted over time at the appropriate discount rate for the individual firm. The cumulative discounted cash flow over time then can be given by:

$$262 \qquad CDCF(t) = \sum_{t=1}^{T} \frac{CF(t)}{\left[1 + \alpha\,(t-t-1)\right]^{t}}$$

where   is the discount factor, and $(t-t-1) = T = 1$, or we discount over every separate period, α can be defined as the interest rate or the firm's rate of return, that is, the rate at which the cumulative discounted cash flow at the end of the tth time period will be equal to zero. This will occur at the end of the lease or at the end of economic production for the reservoir whichever occurs first. For various levels of α, Figure 39 shows the behavior of the cumulative discounted cash flow for various levels of the discount rate.

The problem may now be defined as follows: it is required to find and compute the schedule of drilling and rig purchases and the schedule of when and how many wells to drill over time such that the rate of return for the firm

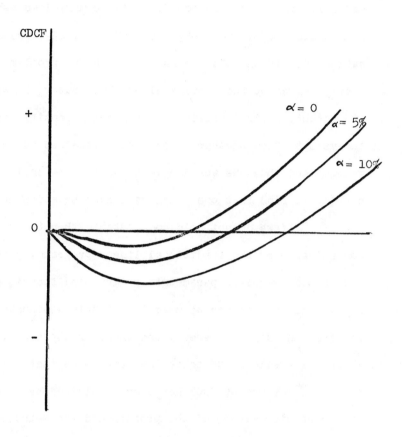

Figure 39   Cumulative Discounted Cash Flow

292

over time will be an optimum, that is, such that the objective function Equation 262 will be maximized subject to Equations 253, 254, 255, 256, 257, 258, 259, and 260 are satisfied. If we hold   constant, then the problem is easily solved by the Simplex Algorithm. But while we hold αconstant, it is impossible to perform a sensitivity analysis to know whether or not the α obtained is the optimum α. And thus, we would have to search the entire arrays of α's to find that one which optimizes Equation 262.

If we do not hold α constant, then the problem is not solvable by the traditional linear programming technique but instead requires parametric programming techniques. The PLP technique that can be used is not unique or unknown but is given in many references, and there exists computer codes which are available to solve the problem as stated above.[2]

The use of CDCF has been criticized because of the time consuming nature of the problem and the methodology of trial and error used to obtain the ultimate solution. Also, it has been taken under scrutiny for its inability to adopt to cash flow streams yielding multiple return solutions, and

[2]See for example; Saul I. Gass, Linear Programming, Methods and Applications, Chapters 8 and 9, (New York, McGraw-Hill, 1958), for specific PLP techniques.

further as not being a useful instrument in selecting between mutually exclusive investment alternatives.

On the other hand, discounted net profit (DNP) is assumed in the economic literature to be a more reliable measure, since it discounts at the average opportunity cost of the individual firm, or at the cost of capital for the individual firm, provided of course that the cost of capital has been defined for the firm. Another possible measure aside from DNP is percentage return on investment PRI. This measures the gain an investment is expected to make over capital invested in the average opportunity. PRI likewise considers the possibility that any investment project may require re-investmnet and thus allows for this possibility. PRI can be given mathematical expression as in Equation

263
$$PRI = \frac{(1+\alpha)^t\,(\alpha)}{(1-\alpha)^T - 1}\left[-C_1 + \frac{R_1}{(1+\alpha)^1} + \frac{R_2}{(1+\alpha)^2} - \ldots\ldots \frac{R_t}{(1+\alpha)^t}\right]\,/C$$

PRI provides then a measure of the average risk. Also, it considers the re-investment potential at the average opportunity rate for the individual firm when considering projects and thus serves as a good project evaluator since the basis of comparison is project return to the standard of average

investment opportunity. The measure of investment worth
should be one that provides the difference between investing in a given project and investing in the average alternative project. The difference can be measured in net present
worth, or as a percent of annual return on investment, PRI.
In the remainder of this paper we shall measure it in terms
of net present worth, but we should realize that there are
other possible methods which could serve as measures of this
difference.

In conclusion, it would be possible to expand the
above linear model to cover for the individual firm all
possible drilling schedules and operations of rigs and thus
the entire array of well spacings available to the firm in a
given time period. This then would yield as an objective
function some concept of return, either in terms of cumulative cash flow, discounted cash flow, net discounted profit,
or some other such term which would be maximized over the
entire set of investment alternatives which the firm had
available. We would then have some optimum investment
policy for the firm to follow, the optimum number of wells
which the firm should invest in, and the production facilities for the firm either on an individual reservoir or
firm basis.

Dynamic Models

    In setting up the dynamic model for the timing and scheduling of wells and the timing and scheduling of investment in productive facilities for a given reservoir, we shall again make use of many of the assumptions made in the linear programming model. However, before organizing the problem in its mathematical form, it is useful to examine the assumptions, the characteristics, and the organization within which the dynamic model will be developed.

    The first set of assumptions are concerned with are those imposed by the rigs and drilling scheduling. These are the same as has been for the linear model so the reader is referred back to that section. We start here then with the added individual assumptions which the dynamic model necessitates. The demand for the homogeneous product is based on past sales, and we add the assumption that it is expected to grow over time, either linearly, expontially, so some other growth curve is followed. Investment is then required for two purposes: first, to meet the expanding market demand, and this will be determined by the firm's own commitment to maintain a position in that expanding market; and secondly, as production continues over time, investment in wells and facilities are necessary to maintain the rate

of production from the reservoir as pressure declines regardless of pressure maintenance. This re-investment process --
the timing and scheduling of the investment projects --
occurs when economic ultimate recovery is still possible and
new investment is feasible. The number of new projects that
become feasible will decline after the initial set of investment projects are completed and as the pressure declines
over time, and as the reservoir nears shut-in or secondary
recovery.

It is assumed that the firm has a number of possible investment alternatives, each of which expands production by a known amount, and the expansion in capacity is
achieved at a known cost since the cost of prior wells is
known. Each of these investment alternatives requires a
significant commitment of funds. Investment in oil and gas
production facilities alone in the United States has been
consistently near 50% of the total of investment allocations
by United States firms in the oil and gas industry, which if
translated in dollar terms is in the two billion dollar
class. It is assumed further that the funds available for
productive investment projects are limited, and thus, for
every investment project there is an opportunity cost in
terms of alternative investment projects the firm could have

invested in at any given point in time. It is assumed
further that the firm obtains the funds from either retained
earnings, or through the issuance of some kind of corporate
debt. Lastly, the firm's planning horizon is finite and is
known, and extends from time zero when the reservoir is
discovered to time period t when it is shut-in or is given
over to secondary recovery.

Given this structure and framework, the problem
facing the firm is then to drill enough wells so as to maximize the flow rate of product such that a maximum net
earnings-ante or post-tax earnings is obtained. Thus, we
are seeking that set of investment alternatives which will
satisfy: (1) the rig and drilling constraints; (2) the
production constraint; (3) the demand constraint for any
period is satisfied; and (4) such that the firm stays within
the funds available at every period for investment purposes.

In the section on the linear model, we exposed
some of the various forms the objective function could take.
In the model we are now considering, we will define the
earnings function as in Equation 261 and 263 and discount it
over time. However, the function we are now concerned with
is an after-tax earnings function so that it is possible to
realize what tax contributes to the corporate earnings

function via the depletion allowance. The depletion allowance is necessarily considered since it is believed to add significantly to earnings, and thus affects the amount of funds that the firm will have available for investment purposes. It is necessary also to include depreciation allowance since the gas-oil industry is capital intensive, and since it is believed that the depreciation allowance will likewise affect the after-tax earnings. Thus, both depletion and depreciation represent sources of funds for the firm for future investment expenditures, and directly add to corporate equity and/or represent a source of funds for retiring the corporate debt. The inclusion of these is essential then when considering the individual firm or reservoir.

The first set of constraints are those dealing with the actual drilling of wells. First, in any period of of time, it is impossible to have more wells drilled than there exists rigs to drill them, assuming that one rig can drill more than one well on any reservoir in any given time period. Likewise, at any point in time, the number of wells drilled up to that point in time cannot exceed the maximum number of wells that have geological and geographical possibility.

Assuming that we have determined the number and the schedule of new wells, and assuming that every well drilled follows the production decline curve postulated, then the production for each well times the number of wells drilled must not exceed the pipeline capacity for the reservoir at any given point in time. The sum over the life of the reservoir represents the reserves taken out, and likewise this cannot exceed the percent of reserves that are actually recoverable from any given reservoir. For the initial calculation of these reserves, the reader is referred to Chapter 3 and Chapter 4.

The third set of constraints imposed on the firm by the problem is that the firm remain within the demand for the product. This restriction is imposed here to approximate the situation as it exists in the largest producing states, where production for each reservoir is given as a percent of the market demand. The demand restriction depends on: (1) the growth of demand over time; and (2) the firm's market intentions of sharing or not sharing in the growth of this market over time. This will obviously depend on the individual firms own decision process. However, for efficiency, it is assumed that no waste occurs, and thus for any given reservoir, the firm will operate within the demand restriction.

Lastly, since the investment in productive facilities, rigs, drilling, and completion materials requires
money, the availability of funds for a given firm imposes
the last restriction. If the firm has no debt ceiling,
and/or if the firm has an infinite supply of funds on hand,
then no financial restriction need be imposed on the firm's
activities. However, even with regard to the largest firms
in the industry, this assumption would not hold since the
amount of funds for investment purposes in any firm is
finite, and for every investment project available to the
firm, there exists an opportunity costs in terms of funds
allocated in alternative investment projects. Thus, we
postulate that the funds available for investment purposes
are a function of the firm's present asset position as well
as the firm's current debt ceiling, assuming that some debt
ceiling exists for every firm. Further, as the firm
increases in size, implicitly it is assumed that the earnings
will increase, as well as the firm's assets and permissable
debt ceiling. Thus, the funds available for investment purposes will increase, and likewise the firm's permitable debt
ceiling will increase over time and as the firm grows. The
implicit assumption made throughout the analysis is that the

firm will fund investment projects either through retained earnings (include depreciation and depletion allowances) and through debt financing, which in the final analysis may include equity financing.

Mathematical Formulation of Dynamic Model

The first problem necessary to consider is that of determining the optimal well drilling schedule for a given reservoir over time so that the proper number of wells will be drilled. In order to specify the problem, the following functions and variables need to be defined:

t = an index of periods starting with time equal to zero and ending with time T which is the end of the reservoir's life;

j = an index of possible (geological and geographic) wells; it is assumed that $j = 1........N$ possibilities exist on any reservoir;

V_{jt} = the oil and gas in place for each well drilled at time t;

ϕt = the porosity at time t;

h_t = the net pay thickness at time t;

S_{wjt} = interstitial water saturation at time t assuming water drive;

B_{ojt} = oil formation volume factor at time t in bbls/bbls;

A_{jt} = area drainage for each unit well drilled in feet or acres;

A_{jmin} = minimum area drainage, usually required by statute;

A_{jmax} = maximum area of drainage.

Then the first subproblem can be formulated as:[3]

264
$$\underset{A_{jt}}{\text{Max}} \quad V_j$$

subject to:

265
$$A_{jmin} \leq A_{jt} \leq A_{jmax}$$

and subject to:

$$\sum_{j=1}^{N} n_{jt} \leq a \sum_{k=1}^{K} p_{kt}$$

$$\sum_{j=1}^{N} n_{jt} \, d(t) \leq R_t$$

$$\sum_{j=1}^{N} n_{jt} \, d(t) < W_c$$

where n_{jt} can be defined as the number of wells drilled in time period t. The number of wells, n_{jt}, is merely a function of A_{jt} since A_{jt} represents the area of drainage per well over the total acreage, which ultimately yields us the number of wells in terms of acres per well.

---

[3]Here V is defined as: $V_{jt} = \dfrac{7{,}758 \; A_t h_t \, \phi_t (1 - S_{wjt})}{B_{ojt}}$

and where 7,758 is the conversion factor from ft^3 to barrels per acre foot.

303

The mathematical structure of the model depends on the assumptions made about h_t, ϕ_t, B_{ojt}, and S_{wjt} as a function of the decision variable A_{jt}. Since we have assumed water drive, and since these variables will be known for each time period t, a linear relation is derived. Solutions to this problem will yield optimal A_{jt}^* over all time periods that is, an optimal area of drainage given the problem constraints, and this will in turn yield the optimal number of wells that should be drilled for any given reservoir.

Subject to A_{jt}^* being an optimum, we are able to arrive at the optimum quantity of reserves being available at any time in the reservoir. Subject to these amounts of reserves being recoverable, and subject to these reserves existing at any time period t, it becomes possible to define the optimal decline curve for the reservoir. To formulate the problem, it is necessary to use the following new functions and variables:

V_{oj} = the initial amount of reserves in well j. This is derived from the materials balance equation;

V_{jt} = the amount of recoverable reserves in well j at time period t;

$V_{j,min}$ = the minimum allowable reserved depleted for economic operation of the reservoir.

$V_{j,max}$ = the maximum reserves possible from reservoir j.

D_t - the demand for the product at time period t;

q_{scjt} - the flow rate from well j in time period t assuming standard operating conditions.

Then it is possible to define the second subproblem as:

266
$$\text{Max} \quad q_{scjt}$$
$$V^*_{jt} \quad .$$

subject to:

267
$$V_{j,min} \leq V_{jt} \leq V_{j,max}$$

$$V_{jt} \geq V_{ja}$$

$$V_{jt} \geq D_t$$

The mathematical structure of this model again depends on the assumptions made about the variables p_{wjt}, h_t, k_t, μ_t, and B_{ojt} as functions of the decision variable V . It is again assumed for simplicity that a linear relation exists for the purposes of our analysis. The solution to this problem will yield optimal q^*_{scjt} over all time periods so that the decline curve followed by the reservoir and each well in the reservoir will be optimal.

Given that this optimal q^*_{scjt} exists, the problem now becomes to maximize the firm's profit subject to these prior restrictions. For this problem, it is necessary to define the following variables:

D_t = again the demand at time period t;

q^*_{scjt} = the optimal production rate from well j at time period t;

$q_{scj,min}$ = the minimum productive capacity at well j in time period t;

$q_{scj,max}$ = the maximum productive capacity at well j for t time period;

t = the net after-tax earnings of the firm.

It is then possible to define the problem as maximizing the net after-tax earnings of the firm subject to the rate of production being optimal, and satisfying the demand constraint, or in terms of our programing model, this may be written:

268
$$\underset{q^*_{scjt}}{\text{Max}} \quad \pi_t$$

subject to:

269
$$q_{scj,min} \leq q_{scjt} \leq q_{scj,max}$$

and:
$$q_{scjt} \geq D_t$$

Here the function π_t is defined as the net after-tax earnings of the firm discounted over time, and which are a function of q^*_{scjt}. Thus π_t can be written:

270
$$\pi_t = (1-Tr) \left[R_{jt} - C_{jt} - O_{jt} \right] - (dr - wv - xn)$$

Again, the mathematical structure of the model
depends on the assumptions made about revenues R and costs C
as a function of the rate of output. Here again we have
assumed that the relationship is a linear one for simplicity.
This formulation of the problem will yield us the rate of
production for the individual wells and multiplying this by
the number of wells drilled will yield the production rate
for the entire reservoir.

The new addition to the analysis is the expression
0_{jt} which now consists first of depreciation and secondly, of
depletion. The value of depreciation depends on the method
of computing the allowance, and since the oil-gas industry
is capital intensive, it is worth noting and isolating this
factor. The method of depreciating capital depends on the
company, and thus will vary. The two general classes which
the federal government allows fall into two categories,
straight-line, and accelerated depreciation. The particular
method used by a company is a function solely of the management, and the conditions which exist in the company. In the
oil-gas industry, since it is capital intensive, the accelerated depreciation would be more advantageous to use because
of the faster write-off of capital. The net effect then of
allowing for accelerated depreciation is to increase in the

early years of an investment the allowance for new capital
and hence to increase the net after-tax earnings at the
beginning of the piece of equipment's life when the extra
earnings are needed. Also, since the after-tax earnings are
increased early in the life of the investment, the overall
objective function will be increased due to the time discounting of returns over the life of the investment. This
would have the tendency of accelerating investment spending
at the beginning of the reservoir's life, to accumulate the
depreciation charges and to accelerate the write-off so as
to increase the earnings of the firm when the capital outlay
is needed.

Secondly, in order to present a more realistic view
of the oil-gas industry, it is impossible to leave from consideration the depletion allowance, and thus O_{jt} introduces
the problem of depletion allowance into the objective function. The current tax code states that the depletion allowance can be claimed on a percent of revenue received from
utilizing the resource in question, oil and gas. The code
further states that the allowance must be calculated on each
property separately. Likewise, the allowance is applicable
only if there is a positive net income from the recovery of
oil-gas, and in the maximum it is limited to an amount equal

to 50% of net income.[4] Depletion cannot be deferred from one period to another, but rather can only be claimed in one tax year. It is possible to express these characteristics in mathematical form as:

271
$$\partial_a = \text{Min} \begin{cases} \text{Max} \begin{cases} 0 \\ \partial_2 \ \text{DNI} \end{cases} \\ \partial_1 - \text{DR} \end{cases}$$

where we define:

∂_a = depletion allowance;

∂_1 = depletion allowance coefficient which in oil-gas is presently 20%;

∂_2 = 50%;

DNI = DR - DC;

DR = discounted revenues;

DC = discounted costs.

This defines three areas of production. First, as long as net income from recovery is negative, the maximum depletion allowance obtainable is zero. Secondly, when there is positive net income, the depletion allowance is equal to 50% of the net income. Thirdly, when this amount of net income exceeds the percent of sales, then the 20% allowance

_____

[4]The Federal Tax System, Facts and Figures, Joint Economic Committee United States Congress, U. S. Government Printing Office, (Washington, D. C.), Sections 611-614, pp. 107-118.

comes into effect. Thus, if we were to diagram this, there
would be two critical points: first, where the net income
goes from negative to zero which would then be the breakeven
point; and secondly, the crossover point which would be
where,

272    $\partial_1\%$ of net sales $= \partial_2\%$ of net income.

Integrating these considerations of the depletion
allowance into the previous analysis, the form of the objective function now becomes as in Equation 273.

273
$$\pi_t = \text{Min} \begin{cases} \text{Max} \begin{cases} (1-Tr)\ NI \\ (1-\partial_2 Tr)\ NI \end{cases} \\ 1 - Tr\ NI - \partial_1 Tr - R \end{cases}$$

where everything has been defined as before. Substituting
in for the definition of NI, we have:

274
$$\pi_t = \text{Min} \begin{cases} \text{Max} \begin{cases} (1-Tr)(R-C) \\ (1-\partial_2 Tr)(R-C) \end{cases} \\ 1 - Tr\ (NS-C) - \partial_1\ Tr\ R \end{cases}$$

For convenience, we again separate π_t into three regions.
The first is again below the breakeven point and thus a
minimum level of production takes place if any at all. The
second is again a crossover point where:

275    $\partial_1\%$ of net sales $= \partial_2\%$ of net income.

The third is the point of maximum production. It is then feasible to define the two critical points in terms of Figure 40 as:

276 $\qquad$ $qA = R = C$

$$qB = (1 - \frac{\partial}{\partial 2} Tr) \, DR\text{-}DC = (1\text{-}Tr)(DR\text{-}DC \frac{\partial}{\partial 2} Tr \, DR)$$

so that:

277 $\qquad$ $qB = DR \left(1 - \frac{\partial 1}{\partial 2}\right) = DC$

With these points determined, the critical points can now be determined for any given cost and revenue function. Thus

278 $\qquad$ $\pi_{jt} = (1\text{-}Tr)(\sigma_j q_{scjt} - FT_j)$ if $0 \leq q^*_{scjt} \leq qA_j$

279 $\qquad$ $\pi_{jt} = (1\text{-}Tr)(\sigma_j q^*_{scjt} - FT_j)$ if $qA_j \leq q^*_{scjt} \leq qB_j$

280 $\qquad$ $\pi_{jt} = (1\text{-}Tr)(\sigma_j q^*_{scjt} - FT_r)$ if $qB_j \leq q^*_{scjt} \leq$

$$q_{scj\,max}$$

where we define σ_j as the incremental profit defined from Equations 278, 279, 280.

The effect of the depletion allowance on the objective function is then to increase the net after-tax earnings. By how much can be determined if we perform a sensitivity analysis allowing the depletion allowance to vary from $0 - 27\frac{1}{2}\%$ and view how net after-tax earnings behave. However, the inclusion of the depletion allowance

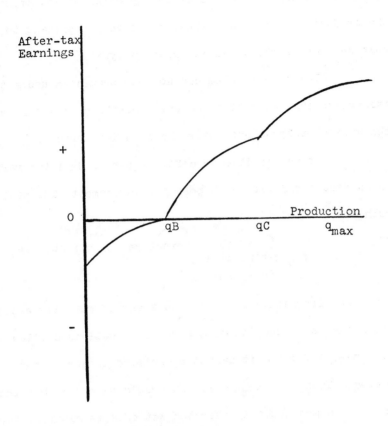

Figure 40   Production and Earnings Function

complicates the problem. The function q now becomes piecewise linear or concave depending on where you are on the production curve, and thus, it is very important to know what segment of the curve is under analysis.

The main problem can now be formulated under the assumption that an optimal solution has been obtained for the various subproblems, over the planning horizon.

It is possible to define a set of decision variables associated with each possible investment alternative such that:

281
$$I_t = \begin{cases} 1 \text{ if the alternative is initiated in} \\ \text{time period t.} \\ 0 \text{ if not.} \end{cases}$$

For each alternative I_{jt} there is a set of mutually exclusive variables as a function of t since if a well is drilled in time period $T = t'$, it cannot be drilled in some other time period. Thus $0 \leq I_{jt} \leq 1$. It should be noted that the index j ranges not only over the set of alternatives, but also over the set of existing facilities at time zero, and thus we can define: $I_{jo} = 1$ for all possible N, and where we assume that at time zero at least one well has been drilled since it was necessary to know that reserves existed on the reservoir. This again is the experimental well.

Another variable a_t is defined as a vector having

as many elements as there are periods in the planning horizon, and there are one's in the tth position and zero's
thereafter. For example, some of the a_t's are define below:

282 $\qquad a_1 = \begin{bmatrix} 1, & 1 & \ldots\ldots & 1 \end{bmatrix}$

$\qquad\quad a_2 = \begin{bmatrix} 0, & 1, & 1 & \ldots\ldots & 1 \end{bmatrix}$

$\qquad\quad a_m = \begin{bmatrix} 0, & 0, & \ldots\ldots & 0 & 1 \end{bmatrix}$

This variable used with I_{jt} is to indicate when expansion of
productive capacity is available, and thus when an increase
in after-tax earnings is possible from drilling the jth additional well.

Then for a given period t and a given set of well
investment alternatives I_{jt}, it is possible to define:

283 $$\sum_{j}^{N} a_t I_{jt}\ \pi_t$$

which is the optimal after-tax earnings given the optimal
investment plan. Summing over all wells drilled and discounting the earnings over time, we obtain the discounted
optimal after-tax earnings for the reservoir given the
optimal investment schedule:

284 $$\sum_{t}^{T} \sum_{j}^{N} \alpha_t I_{jt}\ \pi_t$$

which is the present value of the sum of the optimal after-tax earnings for each period over the planning horizon.

Thus, we define our main objective function as:

285
$$\text{Max}_{I_{jt}} \sum_{t}^{T} \sum_{j}^{N} (\alpha_t I_{jt}) \, \pi_t$$

where α = discount rate. The problem will be maximized by the choice of I_{jt}.

Since the drilling of wells costs money, there will exist at each point in time a constraint in the form of funds available for the investment in new wells. There is also associated with each possible investment alternative an investment parameter. This parameter has as many elements as periods in the planning horizon. It is defined as having a non-zero element starting with the tth period and extends as such for the number of periods in which investment is required. Hence if the ith alternative requires investment B_1, B_2, B_3, to be made in three periods, then B_1 would have the form:

286
$$B_i = \begin{bmatrix} 0, & 0, & \dots & B_1, & B_2, & B_3, & 0 & \dots & 0 \end{bmatrix}$$

It is assumed that the new well is capable of production in the last period when the increment of investment is required. For example, in the above example the well could start production in the $i - 3^{rd}$ period.

We need now to turn our attention to the limitation of funds available to the firm in any given period. The limitation of funds can be stated for any period of time t:

287
$$T_t = \eta \, \mathcal{E}_t - D_t$$

where:

F_t = funds available in period t;

η = debt/equity ratio.

$\mathcal{E}_t$ = equity at time t.

D_t = outstanding debt at time t.

Further, we can express equity at any time t as:

288
$$\mathcal{E}_t = \mathcal{E}_{t-1} - \Delta R_e$$

where:

$\mathcal{E}_o$ = the initial equity at the beginning of the horizon.

$\mathcal{E}_{t-1}$ = the equity at time t-1

R_e = the change in retained earnings from one period to another

K = dividends.

Using the above, we can express equity recursively as:

289
$$\mathcal{E}_t = \mathcal{E}_o + \sum_{t=1}^{T} \pi_t - K\,t$$

Likewise, we can express debt at any point in time as:

290                 $D_t = D_{t-1} + B_{t-1} - \Delta R_e$

where everything has been defined as before only,

B_{t-1} = the total investment in the previous period.

As we can see from the above, we have assumed that investment in wells will be financed either by retained earnings
or by the issuance of debt. We can postulate the debt
relation recursively as:

291          $$D_t = D_0 + \sum_{t=1}^{T} B_t - \sum_{t=1}^{T} \pi_t - K t.$$

Combining these into the original budget constraint, it is
possible to write:

292

$$\sum_{j=1}^{N} I_{jt} \, a_t \, B_{jt} \leq \eta \, \varepsilon_0 + (1+\eta) \sum_{t=1}^{T}$$

$$\pi_t - (1+\eta) \, Kt \sum_{t=0}^{T-1} B_t - D_0$$

Rearranging terms, we can write Equation 292:

293

$$\sum_{j=1}^{n} I_{jt} \, a_t \, B_{jt} + \sum_{t=0}^{T-1} B_t - (1+\eta)$$

$$\sum_{t=1}^{T} \pi_t \leq \eta \, \varepsilon_0 - D_0 - (1+\eta) \, Kt$$

which in words says that:

294   (current investment) - (past investment) -

(past earnings) = (initial loan base) -

(dividends paid).

The necessary steps for the solution of the problem is to find the optimal solutions to the various subproblems, and then to allocate the total production to the existing facilities, and any new alternatives selected for consideration. The objective function is to maximize total after-tax earnings for any given time period while at the same time satisfying the constraints of the problem.

The mathematical structure of the problem depends on the form of the cost and revenue function. We have seen earlier the various forms these functions can take, and we assume for our purposes that both are linear.

The optimization technique applicable in all cases is that of dynamic programming with each well investment or each facility being a stage of the process. The state variable is the amount of production to allocate. Thus, we can let:

q_{scj} = the amount of production that is being allocated;

$q_{scj,max}$ = the maximum capacity at reservoir j that can be allocated;

$q_{scj,min}$ = minimum capacity at reservoir j that can be allocated;

N – the number of investment alternatives that are under consideration;

$\pi_j(q^*_{scj})$ – the profit at reservoir j when q_{sc} is being allocated.

Several restrictions and relationships apply to the problem as it has been formulated above. These restrictions are:

295    $q_{scj,min} \leq q_{scj} \leq q_{scj,max}$

296    $A_j = \text{Max} \left\{ q_{sc} - \sum_{k=1}^{j-1} q_{k,min} \right\}$ is the lower bound production at reservoir j.

297    $B_j = \text{Min} \left\{ q_{sc} - \sum_{k=1}^{j-1} q_{k,max} \right\}$ is the upper bound on production at reservoir j.

and:

298    $F_o(q_{sc}) = 0.$

Then we can set the problem in series form as follows:

299    $F_1(q_{sc}) = \underset{A_1 \leq q_1 \leq B_1}{\text{Max}} \left\{ \pi_1(q_1) \right\}$

$F_2(q_{sc}) = \underset{A_2 \leq q_2 \leq B_2}{\text{Max}} \left\{ \pi_2(q_2) - F_1 (q_{sc} - q_2) \right\}$

.
.
.

$$F_1(q_{sc}) = \underset{A_1 \leqq q_i \leqq B_1}{\text{Max}} \left\{ \overline{\pi}_1(q_1) - F_{i-1} (q_{sc}-q_1) \right\}$$

The structure of this dynamic programming problem is of the type described by Bellman, and of a similar form as described in Chapter 9 of this thesis. The solution posed by Bellman is applicable to the solution of this problem as formulated. We shall see the actual solution process in the next chapter, and see why the problem has been formulated mathematically as here in Chapter 8 and in Chapter 6. Thus, having formulated the problem, it is now necessary to look at the technique of solution.

CHAPTER 9

DYNAMIC SIMULATION AND COMPUTATIONAL

METHODS OF SOLUTION

Introduction

To the present, the main concern has been the
mathematical formulation of the problem in the context of
models, both linear and dynamic. The problem which confronts
us in this chapter will be the presentation of a solution to
the problem as it has been formulated.

First, in regard to the linear models developed in
Chapters 6 and 8, there exists many algorithms and methods
of solution to the problems as formulated therein. The
simplex algorithm, the revised simplex method, and in some
cases parametric techniques, can be used to obtain numerical
solutions.[1] Extension of these models would be feasible,

---

[1]For a treatment of the simplex, revised simplex,
and parametric programming techniques, there exists at this
time many excellent treatments and methods of analysis.
Dantzig, Gass, Hadley, and some other standard references are
contained in the Bibliography of this thesis. See Knuth also
for the computer programming aspects of these algorithms.

extension mainly in the direction of allowing only for integer solutions to the investment and well spacing problem.

However, while these extensions would be challeng-
ing, this section will be concerned with the dynamic models
as formulated. The first problem encountered is the solution to the various subproblems in the relevant chapters.
The methodology to be used is the allocative algorithm
described by Bellman in his treatment of the principle of
optimality.[2] The various subproblems have been formulated
so that first the principle of optimality applies, and
secondly, that the allocative algorithm postulated applies
to the problem as formulated. The principle of optimality
states:

> "An optimal policy has the property that
> whatever the initial state and initial
> decision are, the remaining decisions
> must constitute an optimal policy with
> regard to the state resulting from the
> first decision."[3]

From this statement it is clear that there exists an initial
state and initial conditions, and from this initial situation

---

[2]R. E. Bellman and S. E. Dreyfus, Applied Dynamic
Programming, (Princeton, N. J., Princeton University Press,
1962), Chapter 1, p. 15 and pp. 21-25.

[3]Ibid., p. 15, and also stated originally in R. E.
Bellman, Dynamic Programming, (Princeton, N. J., Princeton
University Press, 1957), p. 83.

other states may evolve. The evolution of these other
states, however, must constitute an optimal decision sequence
with regard to these initial conditions. To refer to the
statement of the problem as it has been formulated in Chapters
6 and 8, suppose the return function, without discounting is
given by π. If the initial state of production is given
by q_0 and the other production states by $q_1\text{---}q_i\text{---}q_m\text{---}q_n$
then

$$\pi = \pi(q_0)$$

This value function could also be defined by Equation 301:

301   $\pi(q_0) = \underset{q_1\cdots\cdots q_N}{\text{Max}} (Fq_0q_1 + Fq_1q_2 + \cdots + Fq_{n-1}q_n)$

where it is assumed that there exists a transition from the
initial state q_0 to the final state q_n. The decision process will exist and will be feasible provided that there
exists a path from q_0 to q_n. Denoting this path by Gq_i,
then for Equation 301, it is possible to write:

302   $\pi(q_0) = \underset{q_1 \quad Gq_i}{\text{Max}} \left[Fq_0q_1 + \underset{q_2\cdots q_n \, \in Gq_i}{\text{Max}} \ Fq_1q_2 + Fq_2q_3 + \cdots \quad Fq_{n-1}q_n \right]$

However, by definition, the second term of Equation 302 is:

$$\underset{q_2\cdots q_n \, \in \, Gq_i}{\text{Max}} \left[Fq_1q_2 + Fq_2q_3 + \cdots Fq_{n-1}q_n \right]$$

which is equal to $\pi(q_1)$.

^{S}o that for any state q_1 of production it is possible to write:

303
$$\pi(q_1) = \text{Max} \quad \left[Fq_0 q_1 - \pi(q_1) \right]$$
$$q_1 \in Gq_1$$

which shows that the principle of optimality is applicable. Reviewing Chapters 6 and 8 finds the problem formulated in this manner.

However, there has been one extension if viewed carefully. Namely, revenues and costs have been discounted over time, and thus, the state function altered somewhat. Now Equation 303 in our system becomes

304
$$\pi_0(q_0) = F(q_0)$$

where $F(q_0)$ can be either some positive return or zero. In the case of Chapters 6 and 8 we have allowed $F(q_0)$ to be zero. However, in either case, $\pi_0(q_0)$ is not discounted, since it is the initial state or initial return. The remainder of the returns must be discounted over the life or horizon of the program. Then for period 1, the decision problem is maximizing:

305
$$\pi_1(q_1) = \text{Max} \quad \left[Fq_0 q_1 - \left(\frac{1}{1+\alpha} \right) \pi_0(q_0) \right]$$
$$q_1 \in Gq_1$$

where α is the appropriate discount rate for the individual firm. For the ith period, then the decision problem takes the form:

306
$$\pi_1(q_1) = \text{Max} \left[Fq_0 q_1 + \frac{1}{(1+\alpha)} \, 1 \right.$$
$$\left. \pi_{1-1} \, (q_1 - 1) \right] .$$

For the nth period then:

307
$$\pi(q_n) = \text{Max} \left[Fq_0 q_n + \frac{1}{(1+\alpha)}^n \; \pi_{n-1} \, (q_{n-1}) \right]$$

where in all cases α is assumed to be constant over the N period horizon. As the principle of optimality was shown to hold in the undiscounted case, so here the principle of optimality will hold in the case of discounting.

In regard to the allocative algorithm, the reference is again to Bellman.[4] The algorithm will not be reproduced here since it is treated in extension in the reference. For the subproblems, the first step is to allocate the total production over time among the facilities that exist in that time period, and once this production is allocated to select any new alternatives for consideration when necessary. The objective function in both cases is to maximize the earnings

---

[4]Ibid., pp. 21-25.

function as it was defined in Chapter 8, while for each period staying within the constraints of the problem. The structure of each subproblem, as has been mentioned, depends on the form of the function, and these forms have been specified for each problem in either Chapter 6 or Chapter 8. Regardless of the form, the optimization techniques applicable is dynamic programming with each facility or well being a stage in the process. Thus, the format is exactly as stated by Bellman and the algorithm is used for solution.

Methodology

In devising a computational scheme to solve the formulated problem, it is necessary to revert back to the recurrence relation given in Equation 299. There the problem is to calculate the elements of the sequence $F_{i-1}(q_{sc} - q_i)$ and likewise the assumption is that these set of q's will be non-negative. It is impossible to tabulate all the values which a function may take, or even any very large finite set of values that the function might take. Thus, it is necessary to use some type of interpolation which will aid in the numberical tabulation of the function. This will yield the amount of production that has to be allocated over each time period, and the amount of investment that is necessary in each time period to achieve the amount of production that is being allocated.

To present the set of values for $F_n(q_{sc})$ over the interval $t = 0, 1, \ldots, T$, it is necessary to make some assumption about the initial set of conditions at $t = 0$. It is assumed then at $t = 0$ the firm has N wells and productive facilities on the particular reservoir in question where N can take on any positive number or zero. In time period one, two decisions are possible for the firm: it can invest in more facilities, or it can do nothing but remain intact. If the proper decision is to do nothing, then the proper switching node will be called the zero or nothing node. If the decision is to invest, then since the firm already has N facilities, it can add one or more to the N already existing there giving it N + 1, N + 2, and it is assumed that it is possible to add up to H facilities. This gives N + 1, N + 2, ...N + H possible alternatives that have to be considered.

In time period two, the same decision process again arises. There is again the alternative of doing nothing and thus switching to the zero node. If again the decision is to invest, then there are at least H - N - 1 possible alternatives that have to be considered, assuming some action has been taken in time period one. However, in time period two, the nodes that were used in time period one are now no longer available, save the zero node. That is, the wells and

facilities that were realized in time period one still exist
and thus are not possibilities or candidates for investment
by the firm any longer. Thus, they are excluded from all
further consideration once they are realized. Thus, if one
well was invested in during time period one, then in time
period two there were N - 1, N - 2, H - N - 1 possibilities
for investing during time period two. Similarly, in time
period three, there will be at least H - N - 2 possibilities
for consideration along with the zero node. This process of
switching continues until all the possible investment alternatives for the reservoir have been considered, or the end
of the horizon is reached.

The process described above is that of a branch
and bound problem and the computational algorithm will be a
variant of this method.[5]

Putting the process discussed into a decision tree
format would give us Figure 41. At "Start" there exists N
well facilities. The process is either to drill no wells or
one well each time period depending on the addition to the

[5]See Norman Agin, "Optimum Seeking With Branch and
Bound," Management Science, (December 1966), Vol. 13, no. 4,
E. L. Laroler and D. E. Wood, "Branch and Bound Methods,"
Journal of the Operations Research Society of America,
(July-August 1966), Vol. 14, no. 4, pp. 699-717, plus many
other articles for the branch and bound method.

328

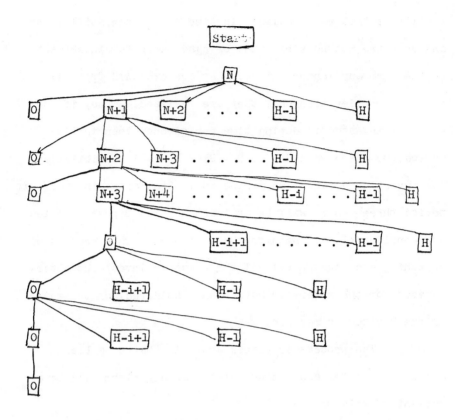

Figure 41   Decision Tree for the Capital
Investment Problem

present value of earnings. As would be expected, the decision to drill no wells becomes the more frequent alternative
as the horizon T is approached or the economic life of the
project is reached.

Computational Scheme

First, it is necessary to assume that the firm will
operate the reservoir at its optimal production rate (MER)
throughout its lifetime. Given this assumption, the problem
is then to find the maximum possible contribution another
investment in a well would make to the return on the reservoir, given of course the life of the reservoir and the life
of the investment. In order to define more clearly the idea
of maximum possible contribution of an investment to the
return for the reservoir, it is necessary to define the following terms and expressions:

U_i = the maximum after-tax earnings for the
     reservoir;

I_i = the investment required for the ith well.
     I_i is a column matrix of the form;

L_i = the economic life of the investment and
     the reservoir since when the reservoir
     is depleted, the investment life is
     concluded.

Since the problem has been formulated as a maximum problem,
it is necessary to be concerned with the upper bound on

earnings, and in particular, the maximum possible contribution an investment can make to after-tax earnings given that
the well investment is operated at optimal production levels
over the economic life of the investment. This upper bound
on the contribution to earnings can be given by:

$$308 \qquad V_1 \text{ max} = \sum_{t=1}^{L1} \frac{U1}{(1+\alpha)^t} - \sum_{t=1}^{L1} \frac{I_{1t}}{(1+\alpha)^t}$$

where:

V_1 max = the upper bound on the contribution to
earnings that another well investment
can make;

α = the discount rate for the individual
firm;

I_{1t} = the column matric defined before

V_1 max then is the upper bound of the present value of an
alternative, namely alternative 1, introduced at time t = 0.
If the same alternative is introduced in any other time
period, say time t = t', then Equation 309 becomes:

$$309 \qquad V_{1t'\text{max}} = \sum_{t=t'+\gamma}^{L_{1t't'}} \frac{U_1}{(1+\alpha)^t} - \sum_{t=t'}^{t'+\gamma} \frac{I_{1t}}{(1+\alpha)^t}$$

But if:

$$\rho = \frac{1}{\alpha} \left[\frac{1}{(1+\alpha)^t} - \frac{1}{(1+\alpha)^t} + L_1 \right]$$

then it is possible to simplify Equation 309

$$310 \qquad V_{it'max} = U_i \rho - \sum_{t=t'}^{t'+\gamma} \frac{I_{it}}{(1+\alpha)^t}$$

Thus, it is possible for each alternative in the decision set
to determine the upper bound on its contribution to the
objective function. The upper bound on this function will
be a function also of the time period it is introduced since
the function is discounted throughout its life.

It can be shown that the upper bound on an investment is a monotonic function of the time period of introduction. As an example of this, consider the case where the
investment cost is incurred for one period. Then Equation 310
becomes:

$$311 \qquad V_{itmax} = \frac{U_i}{\alpha} \left[\frac{1}{(1+\alpha)^t} \quad \frac{1}{(1+\alpha)^t} + L_i \right] \\ - \frac{I_i}{(1+\alpha)^t}$$

If $\dfrac{1}{(1+\alpha)^t}$ is factored out, then Equation 311 becomes:

$$312 \qquad V_{itmax} = \frac{1}{(1\ \alpha)^t} \left[\frac{U_i}{\alpha} - \frac{U_i}{(1\ )^{L_i}} - I_i \right]$$

Now if $\left[\dfrac{U_i}{\alpha} - \dfrac{U_i}{(1+\alpha)^t} - I_i \right]$ is positive, then the upper

bound on the present value of product will be positive and

will decrease over time as $t \to \infty$ since $\dfrac{1}{(1+\alpha)^t}$ becomes

smaller as t becomes larger. If, on the other hand $\left[\dfrac{U_i}{\alpha} - \right.$

$\left. \dfrac{U_i}{(1+\alpha)}L_i - I_i \right]$ is negative then the upper bound on the

present value of product will be negative, but as t increases

over time, $\dfrac{1}{(1+\alpha)^t}$ will decrease, but $\dfrac{1}{(1+\alpha)^t} \left[\dfrac{U_i}{\alpha} - \right.$

$\left. \dfrac{U_i}{(1+\alpha)}L_i - I_i \right]$ will increase as t increases. Thus, the

upper bound on an investment is a monotonic function of the

time period of introduction.

Definitions and Theorems Used for the Branch and
Bound Algorithm

Before considering the normal definitions and

theorems needed for a branch and bound algorithm, it is use-

ful to reflect back on Figure 41 and to view in some detail

what is happening on the decision tree.

First, it is necessary to set the initial condi-

tions of the problem, the start box on the decision tree.

Next, for simplicity it is assumed that only one alternative

can be introduced in each time period. Then having estab-

lished the ground rules, we look at all the nodes in the

decision set for time period $t=1$ and for each alternative

the upper bound on the present value of the objective

function is calculated for each node in the decision set. A comparison is then made of each of these upper bound values, along with the alternative of doing nothing in that period. If there exists a present value of the objective function which is greater than that of introducing no investment at all in that time period, then the node with the highest present value is introduced into the decision set since this node will give the greatest contribution to the present value of the objective function, and thus belongs to the optimal decision set. This node is then considered to remain operable at its optimum level for the remaining time periods until the horizon of the reservoir is reached.

In the next time period t=2, the upper bounds of the present value of the objective function is again calculated. These values are again compared and the alternative of introducing no investment in this time period is again introduced. Again, if there is one alternative which has a maximum upper bound on the present value greater than that of introducing no alternative in this time period, then the alternative with the maximum value is introduced. So for every time period over the planning horizon of the reservoir this process is continued until finally a maximum-maximum is obtained:

313        $U_{max} = V_{1t\ max} + V_{2t\ +\ 1\ max} + V_{3t\ +\ 2\ max}$

$$\cdots\ V_{Nt\ +\ N-1_{max}}$$

where the $V_1 \ldots V_n$ are the upper bound on the present
value of product obtained when the investment alternatives
are introduced in their respective time periods over the
economic life of the reservoir. U_{max} then can be truly
defined as a maximum maximorum for it represents the upper
bound on the present value of the objective functions for
every time period and with the introduction of every possible investment alternative over the economic life of the
project.

Before giving the steps used in the branch and
bound algorithm which will yield the optimal present value
of product and time investment phasing, it is necessary to
define and find the characteristics of the methodology which
will be used. It is useful to keep in mind that the objective function is being maximized subject to the various
constraints mentioned.

Definition 1.   A basic policy is a policy such that
                     the optimal level of production, q_{sci},
                     for the reservoir is obtained.

Definition 2.   A feasible policy is a policy q_{sci}
                     such that the constraints of the pro-
                     blem are satisfied with regard to
                     production, demand, investment and
                     the budget for the firm while remaining
                     at the optimal level of production.

Definition 3. q_1 is some solution to the combinatorial problem (i.e., to the sets of problems as outlined).

Definition 4. $Q : Q = (q)$ is the set of all possible solutions to the combinatorial problem.

Definition 5. $S : S$ is a subset of Q, that is, some collection of solutions to the combinatorial problem.

Definition 6. A partition of Q is an exhaustive division of Q into disjoint subsets $S, \ldots S_N$ with the properties that:

$$S_1 \cup S_2 \ldots \cup S_N = Q$$

and

$$S_i \cap S_j = \emptyset \qquad i \neq j$$

Definition 7. Branching is defined as the process of partitioning a subset S into m disjoint subsets $S_1, S_2, \ldots S_m$ where:

$$S_1 \cup S_2 \ldots \cup S_m = S$$

and

$$S_i \cap S_j = \emptyset \qquad i \neq j$$

Definition 8. A null node is a node from which the feasible action is doing nothing.

Definition 9. An intermediate node is a node from which no branching has yet taken place.

Definition 10. A final node is a node which has been entered into the final solution process and thus is not used again.

Definition 11.   The upper bound on the present value of the objective function is the sum of the final nodes which have been entered into the solution process.

The General Branch and Bound Algorithm

Using the above terminology, it is then possible to define in a general manner the process of a branch and bound algorithm.

A branch and bound algorithm establishes:

1. A set of rules for branching from nodes to new nodes;

2. determining the upper bounds on the present value for the new nodes;

3. chooses a new intermediate node to branch to next;

4. recognized when a node contains only non-feasible or non-optimal solutions; and

5. recognizes when a final solution is obtained.

Theorem 1.

If conditions 1-5 above are satisfied, then the branch and bound algorithm guarantees that an optimal solution to the problem will eventually be obtained.

Proof:

Since Q is finite, the process of branching will eventually yield any given solution S* as a final node, unless stopped previously. That is, the branching process would lead to a partition of Q for which every subset S_1^* obeys by the definition of a partition:

314

$$S_1^* \cup S_2^* \ldots \cup S_n^* = Q$$

$$S_i^* \cap S_j^* = \emptyset \qquad\qquad i \neq j$$

and $S_i^* = S_i^*$. Thus, all possible final nodes may eventually be generated and the optimal solution to the branch and bound problem can be obtained.[6]

Theorem 2.

If there exists an optimal solution to the branch and bound problem via the branch and bound algorithm, then the optimal solution is unique.

Proof:

Let us assume that there exists some solution to the combinatorial problem q_i^* and that this q_i^* is the optimal solution to the combinatorial problem as outlined. Then if there existed some q'n such that q'n $>$ q_i^*, then there exists a branching process Q' such that S' $>$ S* as a final node; that is, there exists a partition of Q' for which every subset S_i' obeys the definition of a partition and yields a

---

[6]See Norman Agin, "Optimum Seeking with Branch and Bound," Management Science, (September 1966), Vol. 13, no. 4, pp. 178-179.

different optimal solution $S_1' = S_1'$. But this is impossible
since S* is the optimal solution generated, and thus q_1^* must
be the optimal solution to the combinatorial problem as formulated. If $q_1^* = q'n$, then they have the same branching process Q and the same set of final nodes, and are thus the
same optimal path.

A Branch and Bound Algorithm

The following branch and bound algorithm is taken
from the sources listed below. The essential difference is
that here the concern is with a maximization process, whereas
in the references listed, the concern is with a minimization
process. Thus, the necessary changes have been made. A
further difference is that the earnings have been discounted
over time, and in the sources listed this is not done. The
process has been listed in steps to facilitate consideration
and the process of branching and bounding to the various
null, intermediate, and final nodes.[7]

Step 1. Set initial conditions of the problem. Define the
economic life of the project; give the set of investments I,
proposed with the investment required I_{it}, and set the upper

[7]See Norman Agin, Ibid., p. 179, also: Gilbert T.
Howard, Optimal Dynamic Investment, (Baltimore, Johns Hopkins
1967); M.J. Beckman, Dynamic Programming of Economic Decisions,
(Providence, R.I., Spreigler, 1968); A. Kaufman and R. Cruiar,
Dynamic Programming, (New York, Academic Press, 1967).

bound on the returns per period from the investment U_1.

Step 2.   Read in the data, and set the parameters of the problem.

Step 3.   Calculate the upper bound of the present value of each alternative as a function of the time period V_{1t}. That is, calculate:

315
$$V_{1t\ max} = \frac{U_1}{\gamma} \left[\frac{1}{(1+\gamma)}t+\gamma - \frac{1}{(1+\gamma)}t+\gamma+L_1 \right]$$
$$- \sum_{t=T}^{t+\gamma} \frac{I_{1t}}{(H_\gamma)T}$$

After computing these values for all the possible alternatives, store these values in a present value matrix.

Step 4.   The index t will denote the number of activities we are considering. Here in this problem t=1.

Step 5.   Now calculate or determine the demand for every period, the total capacity available every period, and the debit limit for every period.

Step 6.   Beginning with period t=1

    a)   determine the set of feasible alternatives;

    b)   with respect to demand,and debt, and capacity, the set of feasible alternatives.

Step 7.   Rank these alternatives according to the upper

bound on the present value of product,

316
$$U_{11} = \sum_{t=1}^{T} \frac{\pi_t}{(1+\gamma)^t} + V_{11\ max} + V_{12\ max}$$

$$+ \cdots V_{i\ N\ max}$$

Step 8. Select the alternative which has the maximum upper bound, say it is:

317
$$U_{1'1} = \max_{i} U_{11}$$

Step 9. Calculate then for i' and the zero alternative a value of the objective function ω:

318
$$\omega_{o1} = \frac{\pi_{11}}{(1+\gamma)} + \sum_{t=1}^{T} \sum_{i}^{N} \frac{\pi_{it}}{(1+\gamma)^t}$$

$$\omega_{1'1} = \sum_{i} \frac{\pi_{it}}{(1+\gamma)^t} \sum_{t=2}^{T} \sum_{i=1}^{N} \frac{\pi_{it}}{(1+\gamma)^t}$$

Step 10. If $\omega_{1'1} > \omega_{o1}$, then branch to node i' and terminate all other nodes for which $\omega_{o1} \not> U_{11}$.

Step 11. Let i' be the node branched to in the previous period. Now at this new node, determine again the demand, the capacity and the debt limits.

Step 12. Now at i' look at alternatives that are (a) feasible with respect to demand for the period, capacity, and debt limits, (b) have not been introduced in a previous period.

Step 13. Now calculate again the upper bound on each node:

319
$$U_{1t} = V_{1't-1} + \sum_{t=1}^{t} \sum_{i=1}^{N} \frac{\pi_T}{(1+\gamma)^t} +$$

$$V_{1t\ max}$$

Step 14. Select that alternative for which:

320
$$U_{1''t} = \max_{1} U_{1t}$$

Step 15. Calculate again for alternative 1" and node 0 a value of the objective function ω:

321
$$\omega_{o1} = V_{1't-1} \sum_{1}^{N} \frac{\pi_{1t}}{(1+\gamma)^t} \sum_{t=t+1}^{T} \sum_{1}^{N} \frac{\pi_{1t}}{(1+\gamma)^t}$$

322
$$\omega_{1''t} = V_{1't-1} \sum_{1}^{N} \frac{\pi_{1t}}{(1+\gamma)^t} \sum_{t=t+1}^{T} \sum_{1}^{N} \frac{\pi_{1t}}{(1+\gamma)^t}$$

Step 16. If $\omega_{1''t} > \omega_{ot}$ branch to node 1" and terminate all other nodes for which $\omega_{1''t} > U_{1t}$.

Step 17. Continue this same process until all alternatives have been introduced or the end of the horizon reached.

Step 18. Back up from node $\omega_{1''t}$ to node $\omega_{1't-1}$ and determine the upper bound on the remainder of the nodes available. If $\omega_{1''t} > U_{1t}$ eliminate these nodes from further consideration.

Step 19. If any nodes remain, calculate:

323

$$\omega_{jt} = V_{1't-1} \sum_{t=1}^{T} \sum_{i=1}^{N} \frac{\pi_{it}}{(1+\gamma)^t}$$

If $\omega_{1''t} > \omega_{jt}$ terminate node j. If $\omega_{jt} > \omega_{1''t}$, then ω_{jt} becomes the new solution.

Step 20. Continue this comparison for all remaining nodes.

Step 21. Now determine the upper bound on all nodes emanating from 1'.

    (a) If $\omega_{1''t} > U_{1t}$, eliminate these nodes.

    (b) If any nodes remain select the node with the max U_{1j} and call it U_{jt}. Now calculate the ω_{jt} for this node.

Step 22. If $\omega_{1''t} > \omega_{ot}$, terminate the ω_{jt} node. If $\omega_{jt} > \omega_{1''t}$, ω_{jt} becomes the new solution.

Step 23. Do this for all remaining nodes. When all nodes have been considered, go to previous time period and repeat process, etc.

Step 24. When t=0, terminate all procedure and the current $\omega_{1''t}$ is the maximum value that the objective function can take and a solution has been reached.

343

In Figure 42 the procedure is given in flow chart form.

Theorem 3

The algorithm as stated above leads to the optimal solution to the investment problem as stated in Chapters 6 and 8 earlier.

Proof:

In order to prove this theorem, it is necessary to show that: (1) there exists only a finite number of possible paths to the various nodes; (2) that all possible paths are considered and are either brought into the solution or eliminated; (3) and that no path is considered more than once.

First, let us consider the number of paths between nodes. For the problem considered, there exists a finite number of alternatives H - N, plus the zero node in each time period t, and the time period is likewise finite. From the zero node can emanate H-N+1 nodes in the second period. In each succeeding period the number of nodes that can emanate from the zero node are H-N+1, while from any other node the number of nodes that can emanate can be given by H-N+1 + γ where γ is the number of nodes that have been entered into the optimal path at that time period. Since

344

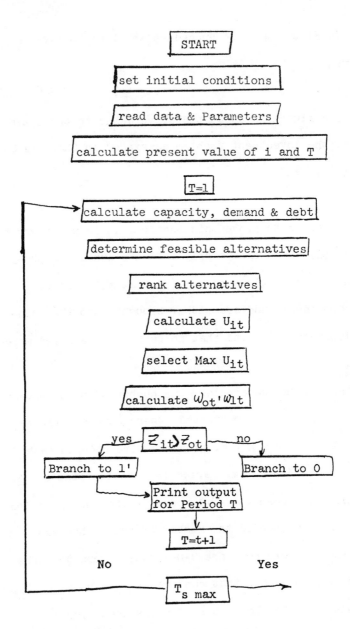

Figure 42 Flow Chart for the Computational
Method Employed

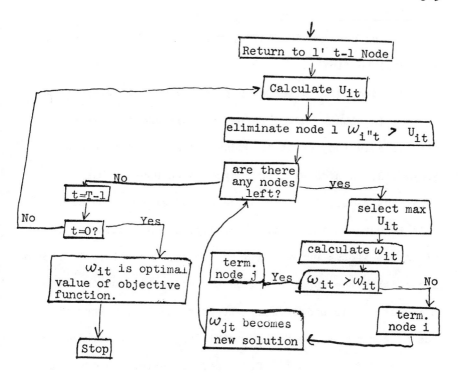

Figure 42   Flow Chart for the Computational
Method Employed

the planning horizon is finite, the number of paths to the
nodes will be finite.

Secondly, after each path and node is considered,
the node is either judged to be feasible and thus entered
into the solution, or it is judged to be infeasible and thus
terminated. A node will be judged to be nonfeasible if the
investment required in that time period exceeds the debt
limit imposed on the firm. Secondly, a node would be judged
infeasible if the upper bound on the present value of product
is less than the upper bound on the objective function when
no investment is undertaken. If this developes, then the
node is considered to be infeasible and is thus terminated.

Thirdly, when a final node is entered into solution, then any other intermediate nodes not brought into the
solution are tested. If the value of their upper bound on
the objective function is greater than the final node, then
the final node is terminated and the new intermediate node
becomes a final node and part of the final solution. This
process continues until all nodes have either been considered
and thus terminated, or they are brought into final solution.
When this is done then all possible paths have been considered and the upper bound on the objective function has
been obtained for the problem.

Method of Calculation

A computer program was written for the algorithm described previously. Here we would like to describe the process used to obtain solutions, and leave the actual solutions to the final chapter on Results and Conclusions, Chapter 10. It will aid the reader to refer back to the flow chart in Figure 42 where the stepwise process has been colligated to its major components and decisions.

The program can be characterized by the following list of traits:

(a) the initial conditions;

(b) the set of possible alternatives;

(c) the revenue and cost functions;

(d) the inclusion of the depletion allowance or the lack of the depletion allowance;

(e) the demand function over time;

(f) the firm dividend policy over the life of the project.

The initial conditions of the program consist of the assumptions surrounding the reservoir at time zero. Factors which needed to be specified were the demand as an increasing function of time, average selling price for the product, initial debt and equity for the firm, the discount

rate the firm used, the prevailing interest rate, the corporate tax rate, and sales and general administrative costs.

Secondly, assumptions had to be made regarding the
set of alternatives available for investment. The set of
existing facilities needed to be specified in order to determine the demand at time $t=0$, and then once the existing
facilities were specified, minimum and maximum capacities
for the reservoir were then able to be determined. Given
the set of existing facilities, then fixed and variable
costs were determined, and these fixed and variable costs
were assumed to hold for the set of investment alternatives
considered. Since a cost figure was determined from the set
of existing facilities, it was assumed that the investment
required in new wells was the same as the cost incurred in
drilling the existing wells. Thus, the investment required
for new facilities was given. And lastly, to obtain revenue,
it was necessary to assume some effective average selling
price for the product.

Having specified the revenue and cost functions,
it was then possible to either include or exclude depletion
from the analysis. Likewise, a sensitivity analysis of
depletion allowance could be performed by allowing the
depletion factor to vary from zero to $27\frac{1}{2}$ percent. Also a

sensitivity analysis could performed by allowing for the rates of interest, and the discount rate to vary over acceptable ranges. For the interest rate the range could be from 4-10%, which the discount rate was allowed to vary from 4-15%. Another possibility was to allow for changes in the effective corporate tax rate by allowing taxes to vary from 40-50% of corporate income.

Another variable which the firm can control and allow to vary is the dividend rate. This parameter could be varied to see what effect various dividend policies would have on the firm. For example, if the percentage of payout is increased, the debt level may be decreased to the point where worthwhile investments might be deferred and thus affect the firm's returns and growth potential. And vice versa if the payout is decreased.

It would be possible to vary the average selling price to see the effects changes in prices would have on the firm's earnings and on the firm's investment. One possibility of price decrease would be if oil import quota were dropped and prices of domestic crude allowed to decline to the world market price. The effects that this decrease in price would have on domestic investment and on domestic returns would then be viewed.

Through the use of this model then the firm could not only gain in knowledge of its own investment alternatives but also the sensitivity of its operation to the various factors and parameters that have been specified.

The CDC6500 was used to run these various tests and changes in parameters. The results of those runs and changes in parameters are given in the Appendices B-J. The results and conclusions and possible extensions of these results are likewise contained in the final Chapter 10.

CHAPTER 10

RESULTS RECOMMENDATIONS AND CONCLUSIONS

Introduction

This final chapter has been divided into three
sections; first, we report the results of the various computer runs made on the model. Secondly, we arrive at conclusions that can be derived from a study of this nature.
Thirdly, we recommend possible extensions to the analysis
undertaken in the body of the thesis, and areas of future
research for those interested in research of this nature.

The results of the computer program that was
simulated for the here presented are contained in
Appendices B-J. The first Appendix B contains the model
as we have initially assumed it. The remainder of the
Appendices contain variations on the model as we have
assumed it. That is the parameters have been varied in
some way to illustrate the effects that the change in the
parameter would have on the reservoir system and investment
patterns.

The changes allowed were in the depletion allowance to see what effect the change in the allowance would
have on the investment program for a firm. Likewise, for
the same purpose changes were allowed in the interest rate,
the discount rate, to see what effects changes in these
parameters would have on the optimal investment policy for
the firm. Taxes were also allowed to vary as well as the
dividend policy the firm might follow. Again we were able
to see the effects changes in these parameters would have on
the investment policy of the firm. Variance in the firm's
debt/equity ratio along with changes in the pricing policy
were also considered. Finally, changes in demand, and in
the initial demand were allowed to see what effect demand
had on the optimal investment policy.

In the second section we are concerned with the
conclusions that can be drawn from a study of this nature.
The conclusions are dependent on the formulation of the
model, and are made in reference to economic and management
theory.

In the third section we are concerned with possible extensions of this thesis, and also related directions
in which future research should go. The recommendations are
in terms of possible theoretical extensions, and secondly in

terms of possible modifications of the model presented
herein, so that more meaningful results and decision processes may be made more realistic for the decision-makers
in the real world.

Use of the Model

The dynamic models developed previously have been
flow charted in Chapter 9 Figure 42 and programmed in
Fortran VI and run on the CDC 6500 version February Mace
for a number of hypothetical problems.[1] In this section we
would like to explain the general format of the program and
the output obtainable from the use of the program, and in
Appendices B-J allow for modifications of the various parameters and inputs the program utilizes.

The program is characterized by the following:

(a) Stored parameters -- the set of
     initial conditions;

(b) A set of investment alternatives
     available to the firm;

(c) Some form of the revenue and
     cost function;

---

[1]For a copy of the program, it is available upon
request to the author.

(d) The inclusion of the depletion allowance in the earnings function and straight-line depreciation being assumed for the depreciation allowance;

(e) Demand-time function and changes in the demand-time relation;

(f) Dividend policy;

(g) Changes in the tax scale applicable to the firm;

(h) Changes in the average selling price.

The initial conditions we assumed were as follows: first, we assumed that the depletion allowance was $.27\frac{1}{2}\%$; secondly, we assumed that the dividend policy for the firm was $.10\%$, and that the allowable debt equity ratio was likewise $.10\%$. These parameters were scaled to be somewhat representative of the relationships as they exist in a "real corporation." Also, we assumed that the interest rate was $.08\%$ and that the applicable discount rate for the firm was $.10\%$. For all the runs made, the initial set of conditions allowed for four existing wells and a set of four alternative investments. While this might be somewhat limiting, the dimension statements are programmed such that they will allow for any number of existing facilities, and

355

any number of investment alternatives. The only program
changes necessary are changes in the H-field for the
appropriate format statements.

The initial demand was such that it was set at
100% of existing capacity. Thus investment in new facilities was required in the first time period to net the
increase in demand. This was done purposely so that we
could view immediately the introduction of investment
alternatives. For each existing facility and alternative
the following factors were immediately read into the computer: the maximum productive capacity; the minimum productive capacity which in all cases was assumed to be zero;
fixed costs for each facility; variable cost for each
facility; the average selling price for the firm to the
pipeline; and the investment required for the initiation of
each investment alternative. The investment requirement
over time was assumed to be increasing for the investment
alternatives and the figures used for each facility were as
follows:

Table 7-Investment Cost for a Set of Investment Alternatives

Facility	Investment Required
1	325
2	375
3	375
4	400

The global parameters were demand as in increasing function of time, which we assumed to be 200 units initially and growing at the rate of 6 units per period, average selling price which we assumed to be $9.00 per unit, general sales, behaviorial, and administrative costs which was assumed to be $25 per period, the tax rate which was assumed to be .50%, the interest rate which was .08% and the discount rate which was .10%. The initial equity for the firm was assumed to be 1,000 and the initial debt was assumed to be 1,250.

The first output we obtained was the upper bound on the present value of product; that is for each investment alternative for each period we calculated the upper bound on the present value of product and the lower bound on the present value of product if that investment alternative could fully recover all the product at that point in time. The upper bound for the initial set of conditions is given in Table 8.

The lower bound on the present value of product was assumed to be zero in that none of the investment projects were assumed to be brought into solution in any period and thus their contribution to product and present value of product was zero.

Table 8- -Upper Bound on the Present Value of Product
(Discount Rate = .10)

Facility	1	2	3	4
Time				
1	824890	824845	824845	824822
2	749900	749859	749859	749838
3	681727	681690	681690	681671
4	619742	619718	619718	619701
5	563411	563380	563380	563364
6	512192	512164	512164	512149
7	465629	465603	465603	465590
8	423299	423276	423276	423264
9	384817	384796	384796	384785
10	349834	349815	349815	349805
11	318031	318013	318013	318005
12	289119	289103	289103	289095
13	262835	262821	262821	262814
14	238941	238928	238928	238922
15	217219	217207	217207	217201
16	197472	197461	197461	197456
17	179520	179510	179510	179505
18	163200	163191	163191	163187
19	148364	148356	148356	148351
20	134876	134869	134869	134865
21	122615	122608	122608	122605
22	111468	111462	111462	111459
23	101334	101329	101329	101326
24	92122	92117	92117	92115
25	83747	83743	83743	83741

The next output received was that of the capital
investment program. For each time period the following
information was received: first the debt, the equity, and
the debt limit; secondly for each investment alternative the
number of units of production for each alternative was first
determined; the net sales in dollars; the production costs
for the units of production; the sales and administrative

costs for that production; the net income from each unit added; the net sales from the reservoir; the net income from the reservoir; the depletion allowance; the taxable income; the tax paid; and lastly the earnings that each facility made in each time period to the firm. Next, the amount of dividends paid by the firm for that time period were calculated and the present value of those dividends over the life of the reservoir were calculated. And finally, the cumulative present value of the dividends paid to date was calculated.

The last output received was the present value of the after-tax earnings for each period and the cumulative present value of the after-tax earnings to date are given in the output. An entire set of solutions to the capital investment program is given in the Appendix J which proved to be one of the more interesting cases for the entire twenty-five time periods. For the format of the capital investment program the reader is referred to that section.

The use of the model is not only for possible decision making for investment, but also to investigate the sensitivity of the various parameters and assumptions on earnings and on the investment decision. For the analysis of the sensitivity of the various parameters the reader is

referred to the Appendices B-J. We would now like to concentrate on the conclusions and the possible extensions to
this thesis.

Conclusions

First, a series of linear and dynamic models for
capital investment in the oil-gas industry have been presented. Specifically we have sought to discuss the problem
of well spacing, however, as the analysis indicates the
models can be extended to cover any capital investment project undertaken by the firm or the industry, only the
specific variables of the models would need to be changed
slightly. The linear models were presented to be used as a
prototype of the dynamic models which were our major concern.

We began by considering the problem in the historical context, namely the historical problem of well
spacing. It was seen that the traditional analysis was
concerned more with the physical aspects of the problem at
first, and only gradually as the physical engineering
principles came to be known as the economic aspect of the
problem recognized. Finally, in the post-war world the
problem was analyzed in its economic context, that is in
terms of discounting revenues and costs, with the initial

capital investment being considered in the cost analysis.

However, while the physical relationships are not the sole determining factors in well spacing, they do enter and thus have to be considered when talking about well spacing or investment in wells. The physical principles enter in two ways; first, in the initial calculation of reserves, and secondly, in the principles of flow of a fluid through a porous media. These two physical problems have to be considered when considering the operation of a well, and thus we have considered these problems in Chapter 3. Thus, having considered these factors it is then possible to obtain some idea of the factors that enter into the determination of a maximum efficient rate of production from a physical standpoint.

However interesting though the physical aspects be, we are more concerned with the economics of the problem, thus, in the flow revenues of the product and the costs involved in obtaining the product. First we present a simple diagramatic presentation of the flows of revenues and costs over time. Relating these to an economic maximum efficient rate of production, it is possible to determine the optimal number of wells that need to be drilled to obtain this optimal flow of revenues and product. However,

the diagramatic methodology is found wanting, and the real
problems that exist in the industry and in the industrial
practice are not contained in the model.

To encompass some of these constraining factors
that in fact exist in reality, it is necessary to work
outside the framework of diagrams, and thus we begin to
explore models in linear and dynamic formulation. We first
begin with linear and dynamic models of production and then
advance to linear and dynamic models of production and
investment. We begin by considering a single well and a
single reservoir and advance to the more complex model of
multi-well multi-reservoir models with well interference.
The restraint of the model take the form of the reserves
themselves, the pressure, and the transportation of the
product. These factors are then incorporated into a dynamic
model whereby we treat the problem in stages and thus
determine the optimal well spacing configuration.

The form of the investment models is similar in
nature, only now along with the production constraints we
introduce the financial constraints under which the firm
must operate. Thus we need to introduce debt and equity
constraints into the model. After setting the model in the
framework of a series of problems, we find that it is

possible to determine the optimal amount and the time-phasing of investment projects for the firm over the life of any investment project.

Having formulated the mathematical framework of the problem, it is then necessary to find a practical solution method whereby it would be possible to determine the investment and the time-phasing of the investment in the project(s). Thus, in Chapter 9 a branch-bound algorithm is presented a one possible method of solving the problem as it was formulated. The algorithm will enter into solution those projects which will yield the highest present value of product discounted over the life of the project, and leave out those which are sub-optimal.

In section B-J we have seen some of the possible changes the model can take on. Since it is difficult to perform sensitivity analysis in dynamic programming, we have allowed some of the easier parameters and variables to vary over the program. It is not exactly sensitivity analysis, however it does show us the sensitivity of changes in some of the parameters to the optimal investment program.

We have shown that dynamic programming is a useful tool to the analysis of economic and business problems. The method allows for considerable realism, and it is possible to obtain value functions and then to compare the alternative value functions of various decisions, or decisions rules. The dynamic programming method has not been used extensively in economics first because of the difficulties of setting the problem in the dynamic programming framework. Secondly, once in this framework, it is necessary to go about finding a solution to the problem that has been analyzed,for no general method of solution exists and thus every problem is unique and requires its own unique solution.

However, because it is difficult is not sufficient reason why it should not be used by economists in the analysis of dynamic problems. Rather, if we are to be concerned with the dynamics of the firm, and if we are to be concerned with the dynamics of the decision making process, we should make use of all the tools that are available, hard as well as easy ones. However, in the future I believe that the technique will be more used for the analysis of dynamic decision making processes such as the capital investment problem we have analyzed here.

Extensions

This study is only a beginning of what can be much fruitful research. The model that was considered

herein was a discrete certainty model. The various parameters and variables were allowed to vary over time and we
have encountered the results of the variations in section
B-J. The first possible and feasible expansion would be in
the realm of uncertainty. Demand, price, etc., the other
variables of the models would each have probability distribution and the implications of this degree of uncertainty or
the after-tax earnings viewed.

A second realm of extensions would be to consider
a continuous model rather than a discrete time dependent
analysis. This would have more possibilities in terms of
theory since it would be an application of control theory to
the realm of economics. V. L. Smith and J. P. Quirk have
used this type of model in analyzing the dynamics of the
fishing industry. It would be possible then to use this
type of model for the oil-gas industry in a special case of
the fishing industry, that is, the fishing industry allowing
for regeneration and oil and gas being fixed stocks with no
generation.

Once set in the framework of control theory, it
would again be possible to extend the certain control
theory results into a theory of uncertainty with again the
variables and parameters having a known probability

distribution function.

In the model we have dealt with in this paper, we have not allowed for economics of scale in investment. It might be possible then to allow investment to be a function of the existing capacity and thus to allow for economics of scale so that it might be possible to write the investment function as:

324 $\qquad$ Capacity $= m(I)^{\propto}$

where $o < \propto < 1$. The problem then becomes whether to invest in the present or to postpone investment to a future period when additional funds are available and the increase in investment would allow for more than a proportional increase in capacity.

Another possibility for future work would be to relax the assumptions we have made in regard to linearity. The firm's objective function might not be linear; likewise, the constraints facing the firm might not be linear. What would happen to the model if these assumptions were relaxed?

In the model we have presented, we have not mentioned inventories. What if we incorporated inventories into the model, and allowed for changes in inventory policy over time for the firm? What would this do to present earnings and what effect would this have on the capital constraints imposed on the firm?

It can be seen then that we have only introduced
an entire class of problems that provide fruitful areas for
both theoretical and applied research.

APPENDICES

APPENDICES

APPENDIX A

The Theory of Dynamic Programming

and Problem Formulation

Dynamic Programming[1]

Introduction. In this section the problem is not
to introduce new methodology to the realm of dynamic programming, but rather to expose the basic elements, definitions and theory, which allow us to formulate and understand
the structure of the problem formulated in Chapters 6, 8
and 9 . Basically we want to understand the mathematical
structure of the recursive programming technique, and from
there be able to formulate the capital investment problem.
Thus, the purpose of this chapter is to introduce the basic
theory of dynamic programming, and secondly, to review the
the post-Smithian literature on the capital investment

---

[1]For the theory of Dynamic Programming there are
many good references. The one we are more dependent on here
is: G. Nemhauser, Introduction to Dynamic Programming,
(New York, John Wiley and Son Inc., 1966). See also the
references in the Bibliography.

problem.

Before embarking on to an exposition of dynamic programming theory, it is necessary and useful to digress on the methodology employed, a methodology which has crept into economics and management gradually. Basically, the idea is the employment of a model, more specifically a mathematical mode. There are an array of definitions of the term "mathematical model," but here it is simply employed to mean a symbolic representation of the relations which exist among the various factors affecting the decision making problem. The basic components of a model are as follows:

(1) Decision. The decision variables are those quantities that can, and will be controlled and manipulated to achieve the desired objective. $D = (d_1, d_2, d_3, \ldots d_n)$.

(2) Parameters. The fixed constants of the system are called parameters. These factors do affect the objective function, but, however, are not controllable in the system but determined outside. $K = (k_1, k_2, k_3, \ldots k_n)$.

(3) Objective Function. The return function is the value associated with the criterion by which the various admissible policies can be judged in the optimization process, i.e., the return with the particular values that the decision variables and the parameters can take on. It is usually

referred to as the profit function, objective function, or occasionally the return function. It has the property that it is a real-valued function of the decision variables and the parameters, and here will be functionally represented as:

325 $$\pi = \pi(D,K)$$

(4) Constraints. The region of feasibility constrains the region within which the decision variables are applicable. These regions are usually referred to as the constraint sets or the region within which the decision variables operate. The feasible values for the decision variables must be within the constraint set. It is usually possible to represent all or part of the constraint set by inequalities of the form:

326 $$F(D) \lessgtr 0 \text{ for all } i$$

(5) Transformation. Each stage of the system transforms the state of its input into an output state in a way dependent on the decisions that have been made for the system at various stages. Thus:

327 $$S_n = T(S_{n+1}; d_n)$$

meaning that given state variable S_{n+1} and decision d_n, it is possible to calculate S_n. $T = (t_1, t_2, t_3, \ldots t_n)$.

(6) State. State variables are those variables which physically characterize the system at any stage. $S = (s_1, s_2, s_3, \ldots s_n)$.

Any D satisfying the constraint set R is known as a feasible solution of the model. The problem is that out of these possible feasible solutions to the problem, which single solution yields the highest profit or return, that is, out of the feasible solutions, which is the "best" possible solution. The "best" solution, call it D*, is defined as that solution which is optimal or satisfied:

328
$$\pi(K) = \pi(D^*,K) \gtreqless \pi(D,K)$$
$$= \underset{D}{\text{Max}} \ \pi(D,K) \text{ where } D \in R.$$

The type of mathematical model used is a determin-istic model. The main characteristic of a deterministic model is that the parameters and the value of the decision variables are specified exactly. There are no uncontrollable variables nor random variables with known or unknown prob-ability distributions. The deterministic model can be con-sidered a special sub-case of the probabilistic model, only in the deterministic model the random variables assume the values with probability one and all other variables with prob-ability zero. Herein, the deterministic model is used because of the comparative ease of manipulation and use which it offered over the probabilistic mode.

Having specified the framework within which the problem is to be formulated, it is then necessary to choose an optimization technique to solve the problem. The method to be employed depends on the objective function, the constraint set, the parameters, and other problem variables. Basically, what dynamic programming does is to transform a sequential or multistage process into a series of single stage decision problems, which, because of the reduction in the number of variables, is an easier problem to handle. The transformation as well as the algorithm used to solve the problem is based on Bellman's Principle of Optimality. More will be said of this optimality criterion later, but it suffices here to say that:

> An optimal policy has the property that whatever the initial state and initial decisions are, the remaining decisions must constitute an optimal policy with regard to the state resulting from the first decision.[2]

Thus a problem with N decision variables may be transformed into N sub-problems, each of which contains only one decision variable.

---

[2]Bellman, R. E., Dynamic Programming, (Princeton, New Jersey, Princeton University Press, 1956), p. 81.

For example, the problem which will concern us later is that of determining an optimal investment policy over time for a reservoir, sequencing the investment decisions so that optimality is maintained with regard to above ground facilities use. Optimality with regard to these latter two problems will yield us an optimal investment sequencing policy as long as optimality is maintained with regard to the first two sub-problems. However, all these decisions are interrelated and optimality with regard to one implies optimality in the others. Thus, whatever the optimal production rate is, this will determine the optimal facilities use and investment phasing will be yielded as a residual. Thus, the optimal production rate serves as an input into the problem of optimal facilities use. Having achieved optimality in facilities use, we then use this as an input to devising an optimal investment and sequencing of a reservoir. This example is merely to illustrate the technique and interrelatedness of the dynamic programming problem which is being solved. Greater detail and exposition to the problem will come in Chapters 6, 8 and 9.

In regard to methods of optimization, the simplest method is the technique of exhaustion or total enumeration. This entails merely the calculation of π (D) for all feasible

D, and then using the definition of optimality to define the set of optimal solutions. However, total enumeration is possible only when there are a finite number of solutions, or if the enumeration be infinite, then it may be possible to estimate by approximating the set of optimal solutions.

Often in dynamic programming it is not feasible to conduct an exhaustive search for the optimum solution in terms of time and economy of the search. If exhaustive search is not feasible, then for attaining the global maximum, a rule or some set of rules must be provided for selecting and evaluating the objective function.

A feasible solution D^0 that satisfies the condition:

329 $$\pi(D^0) = \pi(D^0 + \Delta D)$$

is defined as a local maximum. The set of global maxima is contained in the set of local maxima. Thus, this condition above is necessary for a global optimum to exist. When the local optimum is unique, then the local is the global maximum. Thus, for any function with a unique local optimum, the definition for a local optimum is a necessary and sufficient condition for a global optimum. If there exists more than one local optimum, then there is no choice but to enumerate the optima and conduct a search.

As for search methods, there have been developed a number of methods to evaluate the objective functions. Sequential search procedures, iterative procedures, methods of steepest assent, gradient methods are all ways of moving from one solution to another until the "best" of the various optima has been attained.

The methodology involved in multistage problem solving is to break the problem into its component parts or smaller problems, solve these sub-problems and then combine the solutions to the small problem to obtain a solution to the whole problem. The process of "decomposing" the problem into a series of smaller problems and then "composing" the solution to the smaller problems into the entire system is called multistage problem solving.

To see how this method works, assume that we have a system described by the state variable S_0 and we want to transform this system from state S_0 to state S_N. We assume that there exists some known transformation T_N such that $S_N = T_N(S_0)$. However, assume that there exists some transformation, call it $\mathcal{T}_N$, which transforms a system in state S_{N-1} to state S_N desired. Then $S_N = T_N(S_{N-1})$. Then to go from state S_0 to S_N we need a transformation which will allow us to go from S_0 to S_{N-1}, say T_{N-1}. So our problem, taken in a series, is now composed as follows:

330 $$S_o \to T_{N-1} \to S_{N-1} \to T_N \to S_N.$$

The process can continue, breaking the system into N sub-problems each having the required transformation until the entire N series analysis is described by:

331 $$S_o \to T_1 \to S_1 \to T_2 \to S_2 \ldots T_N \to S_N$$
$$T_{N-1} \to S_{N-1} \ldots T_N \to S_N.$$

The corresponding N sub-problems are also arranged:

332
1. $S_o = T_1 (S_1)$
2. $S_1 = T_2 (S_2)$
3. $S_2 = T_3 (S_3)$
   .
   .
   .
   n $\quad S_{n-1} = T_n (S_n)$
   N $\quad S_{N-1} = T_N (S_N)$

The important fact is that by consideration of the N intermediate states, that is, by breaking the problem into an equivalent series of sub-problems, the entire problem became feasible for solution. There are two methods of analyzing problems, backward multistage analysis and forward multistage analysis. For our purposes we shall assume any multistage problem can be handled equally well by either approach.

A One-Stage Decision System

Thus far the exposition has been concerned with describing and defining multistage systems which are necessary in using and formulating a dynamic programming model. The missing element in what has been described so far is the actual decision making, that is, choosing among the various feasible solutions to the problem. The simplest decision system, the one-state system, can be characterized by the five following factors:

(1) An input stage X that gives the relevant information about the inputs to the problem.

(2) A stage transformation which expresses each component of the output state as a a function of the input state and the decision variables $Y = T(X, D)$.

(3) A decision variable D that controls the operation of the problem.

(4) An output state Y which gives the relevant information about the outputs of the problem.

(5) A stage return ρ which measures the return from the problems as a function of the inputs, decisions and outputs:

$$\rho = \rho(X,D,Y)$$

The single stage initial state optimization problem is to find the maximum stage return as a function of the input stage. Denote F(X) as the optimal and D* = D(X) as the optimal decision policy, then the following holds:

333 $\qquad F(X) = \rho\left[X,D(X)\right] = \rho(X,D^*) = \text{Max } \rho(X,D)$

$$\geqq \rho(X,D).$$

If X and D are both scalars, then the dimensionality of the optimization is one state and one decision variable.

When the optimal return is a function of output Y then a minor complication can occur. If there exists a one to one correspondence between the output state Y and the input state X, then Y = T(X,D) has an inverse which exists and can be given by X = T'(Y,D). Then X can be eliminated from the return function ρ and the stage return can be expressed as a function only of the decisions and outputs:

334 $\qquad \rho = \rho\left[\tau(Y,D)D,Y\right] = \rho(Y,D).$

Then the final state optimization procedure is to choose D as a function of Y so that ρ is maximized. Letting F(Y) this time be the optimal return and again D* = D(Y) be the optimal decision policy, it is possible to express:

335 $\qquad F(Y) = \rho(Y,D^*) = \underset{D}{\text{Max }} \rho(Y,D)$

If it is possible to obtain the inverse as mentioned above, then it is possible to maximize over X and D subject to the stage transformation:

336 $\qquad F(Y) = \underset{DX}{\text{Max }} \rho(X,D,Y)$

$\qquad\qquad$ S.T. $\quad Y = \tau(X,D)$

If the optimal return is determinable by both the input and output states, then it is possible that there may

be no decision making at all. For example, suppose that
$Y = \mathcal{T}(X,D)$ can be solved for D in terms of X and Y, that
is, $D = \hat{\mathcal{T}}(X,Y)$. Then it is possible to express the optimal
return as a function of the input and output states only,
that is:

337 $\qquad F(X,Y) = \rho\left[X \quad \hat{\mathcal{T}}(X,Y)Y\right] = \rho(X,Y)$

The terminology commonly employed in dynamic programming for the optimal stage optimization procedures stated
here are: first, if $F(X)$ is stated as the optimal return,
then it is called the initial state problem, secondly, if
$F(Y)$ is stated as the optimal return then it is a final
state problem; and if the optimal return is determined by
both the input and output states $F(X,Y)$ then it is referred
to as the initial-final state problem.

Serial Multi-Stage Systems

A serial multi-stage system consists of a set of
stages joined together in a series so that the output of
one of the elements in the series becomes the input to the
next element in the series. In general for an n-stage
series where $n = (1...N)$ the generalized stage transformation is given by:

338 $\qquad X_{n-1} = \mathcal{T}_n(X_n D_n),$

and the stage return takes the form:

339 $\qquad \rho_n = \mathcal{T}_n(X_n D_n)$

From the above transformation, it follows that X_n
depends only on the decisions $(d_{n+1}. \;.\; .\; d_n)$ made prior to

stage n and X   itself assuming an initial state problem.
That is:

340
$$X_m = T_{n+1} (X_{n+1}D_{n+1}) = T_{n+1} \left[T_{n+2} (X_{n+2}D_{n+2})D_{n+1} \right]$$

$$= T_{n+1} (X_{n+2}D_{n+2}D_{n+1}) = T_{n+1}\left[T_{n+1}(X_{n+3}\cdot\cdot\right.$$

$$\left. \cdots D_{n+3} \quad D_{n+2}D_{n+1}\right]$$

.

.

.

$$= T_{n+1}(X_n, D_n \cdots D_{n+1})$$

Then from the above AT_n, the return function P_n depends
only on the decisions $(D_{n+1}D_{n+1} \cdots D_n)$ and X_n so the
return function can be expressed:

341
$$P_n = P_n(X_nD_n) = P_n \left[T_{n+1}(X_n, D_n \cdots D_{n+1}) \right.$$

$$\left. D_n \right]$$

$$= P_n(X_n, D_n \cdots D_m)$$

The total return P_n from stages 1 $\ldots$ n are
functions of the individual stage returns and are given by:

342
$$\pi_n(X_N X_{N-1} \cdots X_m \cdots X_1; D_N D_{N-1} \cdots D_m \cdots$$

$$D_1) = g \left[P_N(X_N D_N), P_{N-1}X_{N-1}D_{N-1} \cdots P_m(X_m D_m) \right.$$

$$\left. \cdots P_1(X_1 D_1) \right]$$

And as was mentioned above, the total return depends on all
N decisions $(D_N \cdots D_1)$ and X_N so that total return can
be written as:

343    $\pi_N(X_N D_N \ldots D_1) = g\left[P_N(X_N D_N), P_{N-1} X_N(D_N D_{N-1})\right.$

$\left.\ldots \ldots P_1(X_N D_N \ldots D_1)\right]$

The original optimization problem was to maximize the N-stage return function π_N over the decision variables $D_N \ldots D_1$ as a function of the original input state X_N. If $F_N(X_N)$ is the maximum N-stage return and if $D_m^* = D_m^*$ $(X_N 0$, and $X_m^* = \tau_n(X_N)$ are the optimal decisions and initial states then the generalized maximum N-stage return is given by:

344    $F_N(X_N) = g\left[\rho_N(X_N D_N^*) \quad \rho_{n-1}(X_{N-1}^* D_{N-1}^*) \ldots\right.$

$\left.(X_1^* D_1^*)\right] = \text{Max } g\left[P_N(X_N D_N), P_{N-1}(X_{N-1} D_{N-1}) \ldots\right.$

$\left. \ldots {}_1(X_1 D_1)\right] D_N \ldots D_1$

S.T. $X_{N-1} = \tau_n(X_n D_n)$ $n = 1 \ldots N$.

Decomposition - Additive

The problem is now to take the above generalized form and to decompose it into N equivalent subproblems, each subproblem containing only one state variable N, and one decision variable D. Once having decomposed the problem the N subproblems are then solved, and these solutions are then combined to obtain the solution to the original multistage problem.

In order to achieve this decomposition of the problem, three highly restrictive assumptions have been made with regard to the function g. The first restriction is neither necessary nor sufficient for the problem to be

decomposed, but the remaining two conditions are sufficient
if the problem is to be separated.

The first condition is that of additivity. The
reason this assumption is made is for convenience and mainly
because it would be an unnecessary diversion to dwell on the
problem of generalize decomposition.

By additivity we mean each of the N subproblems
expressed in as a one state variable and decision and thus
equivalent to a one stage optimization problem are interdependent. We find we make one decision at a time and the
solutions combined or added to give the optimal solution.

The second condition for the decomposition of any
function into a series of stages is that the function be
separable. Thus say for example that we have a general
function g of the form:

345 $$g\left[P_N(X_N,D_N), P_{N-1}(X_{N-1}D_{N-1}) \cdots P_1(X_1,D_1)\right]$$

which we are trying to maximize. Then in order to decompose
this N-stage problem into a series of single stages a sufficient condition is that 345 hold:

346 $$g\left[P_N(X_N,D_N), P_{N-1}(X_{N-1},D_{N-1}) \cdots P_1(X_1,D_1)\right]=$$
$$g_1\left[P_N(X_N, D_N), g_2 \ P_{N-1}(X_{N-1},D_{N-1}) \cdots\right.$$
$$\left. P_1(X_1, D_1)\right]$$

where both g_1 and g_2 are real valued functions. And so on,
346 would be separable to $g\left[P_N(X_N,D_N), P_{N-1}(X_{N-1},D_{N-1})\right.$
$$\left. \cdots P_1(X_1,D_1)\right] = g_1\left[P_1(X_N, D_N), g_2 \ P_{N-1}(X_{N-1},D_{N-1}), g_3\right.$$

$P_{N-2}(X_{N-2}, D_{N-2}) \cdots g_N (P_1(X_1, D_1)]$; where $g_1 \cdots$ g_N were all real valued functions.

The third assumption necessary for decomposition of our general problem, and this again is a sufficient condition for decomposing g, is monotonicity. The assumption of monotonicity shows that the sequence of separable, additive functions of g will converge to some uniformly closed bounded set, and thus the limits on g, either upper bounds or lower bounds. Thus, g, is a monotonically non-decreasing function of $g_2 \cdots g$ for every P_N.

Assuming that these three conditions are satisfied then it is possible to write our general function in decomposed form:

347 $\quad g \left[P_N(X_M, D_N), P_{N-1}(X_{N-1}, D_{N-1}) \cdots \right.$

$P_1(X_1, D_1) \right] = P_N(X_N, D_N) + P_{N-1}(X_{N-1}, D_{N-1}) +$

$P_1(X_1, D_1)$

With these assumptions it is now possible to write g as follows:

348 $\quad F_N(X_N) = \text{Max} \left[P_N(X_N, D_N) + P_{N-1}(X_{N-1}, D_{N-1}) + \right.$

$\cdots + P_1(X_1, D_1) \left. \right]; D_N \cdots D_1$

S.T. $X_{n-1} = n(X_n, D_n)$ $\quad$ (n = 1 ... N)

Since the Nth stage return does not depend on the position of maximization, and since for any arbitrary real-valued function, the following condition holds, namely that:

349 $\text{Max}_{u_1 u_2} \left[F_1(u_1) + F_2(u_1 u_2) \right] = \text{Max}_{u_1} \left[F_1(u_1) + \text{Max}_{u_2} F_2(u_1 u_2) \right]$

Then it is possible to rewrite 350 above in the general form:

350 $F_N X_N \; \text{Max}_{D_N} \left[\mathcal{P}_N(X_N, D_N) + \text{Max}_{D_{N-1} \ldots D_1} \mathcal{P}_{N-1}(X_{N-1}, D_{N-1}) + \cdots \right.$

$$\left. \cdots + \mathcal{P}_1(X_1, D_1) \right]$$

S.T. $X_{m-1} = \mathcal{P}_m(X_n, D_n) \quad n = 1 \ldots N.$

From the separability assumption, it is possible to bring the maximum over $D_{N-1} \cdots D_1$ inside the outside parentheses. However, the maximum with respect to D_N is still over $\mathcal{P}_{N-1} \cdots \mathcal{P}_1$, as X_{N-1} depends on D_N through the stage transformation $\mathcal{P}_N$. Then from the definition of maximum N-stage return, $F_N(X_N)$ it follows that:

351 $F_{N-1}(X_{N-1}) = \text{Max}_{D_{N-1} \ldots D_1} \left[\mathcal{P}_{N-1}(X_{N-1}, D_{N-1}) + \cdots + \mathcal{P}_1(X_1, D_1) \right]$

Thus, it is possible to rewrite 351 as follows:

352 $F_N X_N = \text{Max}_{D_N} \left[\mathcal{P}_N(X_N, D_N) + F_{N-1}(X_{N-1}) \right]$

S.T. $X_{n-1} = \mathcal{T}_N(X_N, D_N)$

or:

353 $F_N(X_N) = \text{max} \left[\mathcal{P}_N(X_N, D_N) + F_{N-1} \; \mathcal{P}_N(X_N, D_N) \right]$

However, defining

$\mathcal{M}_N(X_N, D_N) = \mathcal{P}_N(X_N, D_N) + F_{N-1} \; \mathcal{T}_N(X_N, D_N)$

the determination of $F_N(X_N)$ and $D_N^* = D_N(X_N)$ given $F_{N-1}(X_{N-1})$ is simply a single stage initial state optimization problem with state variable X_N, decision variable D_N, and the return function π_N.

Thus the problem, originally an N-stage problem, has been transformed into two simpler optimization problems, namely:

354 $\quad F_{N-1}(X_{N-1}) = \underset{D_{N-1}\ldots D_1}{\text{Max}} \left[\rho_{N-1}(X_{N-1}, D_{N-1}) \ldots \rho_1(X_1, D_1) \right]$

$\quad\quad\quad$ S.T. $\quad X_{N-1} = \tau_N(X_N, D_N)$

and

355 $\quad F_N(X_N) = \underset{D_N}{\text{max}}\ \pi_N(X_N, D_N)$

and together these are equal to equation:

356 $\quad F_N(X_N) = \underset{D_N}{\text{Max}}\ \rho_N(X_N, D_N) + F_{N-1}\ \tau_N(X_N, D_N)$

In fact the process can be continued in the same manner and it is possible to decompose the original N-stage problem into a single stage initial state optimization problem of the form:

357 $\quad F_1(X_1) = \underset{D_1}{\text{Max}}\ \pi_1(D_1, X_1) = \underset{D_1}{\text{Max}}\ \rho_1(X_1, D_1)$

$\quad\quad\quad \cdot$

$\quad\quad\quad \cdot$

$\quad\quad\quad \cdot$

$$F_n(X_n) = \underset{D_n}{\text{Max}} \ \pi_n(X_n,D_n) = \underset{D_n}{\text{Max}} \left[\rho_n(X_n,D_n) + F_{n-1} \right.$$
$$\left. \tau_n(X_n,D_n) \right]$$

.

.

.

$$F_N(X_N) = \underset{D_N}{\text{Max}} \ \pi_N(X_N,D_N) = \underset{D_N}{\text{Max}} \left[\rho_N(X_N,D_N) + F_{N-1} \right.$$
$$\left. \tau_N(X_N,D_N) \right]$$

Terminal Optimization

A special case of the maximization of stage returns is the maximization of some function of the final output state, commonly called terminal optimization. Assume that the problem is now:

358
$$F_N(X_N) = \underset{D_N \ldots D_1}{\text{Max}} \ g(X_0)$$

$$\text{S.T.} \ \ X_{n-1} = \tau_n(X_n,D_n) \quad n = 1 \ldots N$$

Let $\pi_N(X_N,D_N) = \rho_{N-1}(X_{N-1},D_{N-1}) = \rho_2(X_2,D_2) = 0$, and $\rho_1(X_1,D_1) = g\left[\tau_1(X_1,D_1) \right] = g(X_0)$. Then it follows that:

359
$$\rho_N(X_N,D_N) + \rho_{N-1}(X_{N-1},D_{N-1}) + \ldots + \rho_1(X_1,D_1)$$
$$= g(X_0).$$

so that 356 becomes:

360
$$F_N(X_N) = \underset{D_N \ldots D_1}{\text{Max}} \left[\rho_N(X_N,D_N) \ \ldots \ \rho_1(X_1,D_1) \right]$$

$$\text{S.T.} \quad X_{n-1} = \mathcal{T}_n(X_n, D_n) \quad n = 1 \ldots N.$$

It is possible to show that a duality exists between the optimization of the sum of stage returns and the terminal optimization problem since 360 above can be expressed condensely as:

$$361 \qquad F_N(X_N) = \underset{D_N \ldots D_1}{\text{Max}} \sum_{N=1}^{N} \rho_n(X_n, D_n)$$

$$\text{S.T.} \quad X_{n-1} = \mathcal{T}_n(X_n, D_n) \quad n = 1 \ldots N$$

and can be formulated as a terminal optimization problem. Let $Y_N = 0$ and $Y_n = \sum_{K=m+1}^{N} \rho_K(X_K, D_K)$. Then if $Y_o = \sum_{n=1}^{n} \rho_n(X_n, D_n)$ and $Y_{n-1} = Y_n + \rho_n(X_n, D_n)$ $n = 1 \ldots N$. Then 361 can be written:

$$362 \qquad F_N(X_N) = \underset{D_N \ldots D_1}{\text{Max}} \, Y_o$$

$$\text{S.T.} \quad X_{n-1} = \mathcal{T}_n(X_n, D_n)$$

where Y_o indicates that terminal optimization procedures are being used and $Y_{n-1} = Y_n + \rho_n(X_n, D_n)$ $n = 1 \ldots N$. Thus, the required duality is established.

In summary then, it is always possible to transform a terminal optimization problem into a problem of maximizing the sum of stage returns. Likewise, a problem of maximizing the sum of stage returns can be transformed into a terminal optimization problem.

Recursive Equations for Final State and Initial-Final State Optimization

Thus far we have seen in brief detail the initial stage optimization problem, and in the last section the so-called terminal optimization procedures. The third problem of interest is the initial-final state optimization problem. Basically, the initial-final state optimization problem is to find $F_N(X_N, X_0)$, that is, the optimal N-stage return as a function of the input state of stage N and the output state from the initial stage one. There are few differences between this problem and that of the initial state problem. The main difference is that it is not necessary to eliminate X_0 at stage one of the recursive optimization procedure. In fact, at stage one the problem is that of solving a two-stage, single-stage problem of the form:

363
$$F_1(X_1, X_0) = \max_{D_1} \rho_1(X_1, D_1, X_0)$$

$$\text{S.T. } X_0 = \tau_1(X_1, D_1)$$

After determining $F_1(X_1, X_0)$, the remainder of the recursive analysis proceeds as before only at every stage of the analysis X_0 is carried as an additional state. The more general recursive equation is:

364 $$F_n(X_n, X_o) = \underset{D_n}{\text{Max}} \left[P_n(X_n, D_n) + F_{n-1} \cdots \right.$$
$$\left. T_n(X_n, D_n) \quad X_o \right]$$

From a dynamic programming theoretical point of view, the addition of another variable at each stage causes no problem. However, as far as calculations are concerned it can increase considerably the number of state variables.

The final state optimization problem is to find $F_N(X_o)$ the optimal return as a function of the output state from stage one. One way to achieve this is to solve the two-state problem $F_N(X_N, X_o)$, and then to maximize over X. Thus, functionally:

365 $$F_N(X_o) = \underset{X_N}{\text{Max}} F_N(X, X_o)$$

The underlying structure is the direction of the process which is important, important at least some of the time. However, for most multistage decision problems, there is no need to distinguish between initial state versus final state, i.e., forward versus backward recursion. This is so when the choice between input and output is arbitrary from a mathematical point of view. In those cases, it is necessary to construct the transformation so that output states are functions of input states and decisions, and to determine the optimal return as a function of the input state to N-stage using backward recursion.

Decision Trees and Bellman's Principle of Optimality

The topic of the discussion has been that of describing a multistage decision system in which each of the
decisions and the state variables take on a finite array of
values. While this type of system has been described mathematically, it is possible to represent such a system graph-
ically by a decision tree, the circles of the tree (nodes)
corresponding to states, and the arcs corresponding to decisions taken (see Figure 41). Starting at the base of the
tree, there are two possible decisions, each represented by
arcs emanating from the node. So for the second state, the
modes represent the output from stage one which now becomes
input to stage two, each node having arcs or a return emant-
ing from it. So the output to stage two becomes the input
to stage three and the process continues.

The objective of the method is to determine a path
yielding a maximum return. Once the optimal path is obtained,
that is, maximizing the returns from each of the output nodes,
an optimal solution to the system is obtained. There is no
need to consider non-optimal solutions, since they will yield
a non-optimal return.

This process allows us to state Bellman's Principle

of Optimality:

>An optimal policy has the property that
whatever the initial state and decisions are,
the remaining decisions must constitute an
optimal policy with regard to the state
resulting from the first decision.[3]

A proof of the principle need not be stated here since this is not a mathematical treatise. However, a proof could run something as follows: start with the initial state; find one decision that is non-optimal; then the whole policy is non-optimal since the policy can be improved by introducing the optimal policy. Thus, if there are one or more non-optimal decisions, it is possible to improve the entire system by the introduction of optimal policies. The system will be optimal when all the separate decisions are optimal.

_____

[3]Ibid., p. 81 as was given in both of Bellman's works.

APPENDIX B

Changes in Demand

The model as it has been formulated in Chapters 6
and 8 has wide applicability because there are no restrictions placed on the exogeneous variables and the parameters
listed in Chapter 9.

First, there is no restriction that has been
placed on demand for the product as a function of time; any
value that whatsoever might be used in any form. We have
assured demand to be 200 units each time period, but this is
solely for expository purposes. Our demand function is of
the linear form:

366   Demand = Initial Demand + Change in Demand (t-1)
however it could take any other form we might desire it to
take on or any form we might consider interesting. For
example, we might want demand to be an exponential function
of the form:

367   Demand = Initial Demand exp (x (t-1)) where x is
some coeeficient of growth. Another possible demand function

might be of the form

368   Demand = Initial Demand $\exp (B + x (t-1))$ where
B is some initial level of demand and x is the coefficient
of growth.

The general conclusion that we arrived at after
allowing for demand to vary linearly was that demand curves
effect on the objective function was not too pronounced and
hence the solution and the effect on investment and total
earnings are relatively insensitive to variations in the
demand forecast.

Another extension of the model would be to expand
the demand to include the exponential function. Another
variation that could be tested is to apply confidence limits
to estimates of demand and determine the effect of the variations on the investment policy.

APPENDIX C

Changes In Average Selling Price

The second set of changes that were allowed on the model was a change in the average selling price. For the remainder of the analysis, an average selling price of $9.00 per unit or barrel was allowed. However, for variations in the selling price, changes in the selling price were allowed to vary between $3.00 per unit or barrel to $15.00 per unit. Fixed and variable costs were likewise allowed to vary in proportion to the average price change so that competent with each price change was assumed a change in the cost structure. Under these circumstances and under these assumption changes, the optimal investment policy was insensitive to changes in price. Likewise since the cost structure changed with changes in average selling price, the changes in price were insensitive to the objective function, the optimal after-tax earnings function.

Not allowing for changes in the cost structure to change with the change in average price did demonstrate some

sensitivity to the optimal investment policy and to the
after-tax earnings function. The higher the price charged,
the greater the earnings function became, and the greater
the amount of allocated production. The only limiting factor
turned out to be the debt limit. If the debt limit was allowed to vary so that it was no longer a restraint on the
firm, then all the new productive facilities would come into
solution over time as the increase in demand continued.
However, imposing an upper limit on the debt limit caused
fewer of the facilities to be allocated, and thus less sensitivity to changes in price than when debt was not a binding
constraint.

Thus the general conclusion would be that as long
as a debt limit was an upper bound binding constraint, it
was difficult to determine a price effect. However, once
debt was not a binding solution, it was possible to view the
effect which changes in the average price would have on the
optimal investment and optimal after-tax earnings function.

To use then this model to demonstrate what effect
a change in import quotas would have on domestic production,
first we would have to assume the upper limit on debt is
infinite. Then it would be possible to allow price to change
downward seeing the price of crude on the world market is

lower than our supported domestic price, and then to see the
effect this would have on domestic production and investment.
We can demonstrate that less output would be allocated due to
the decrease in price, and secondly, investment in facilities
would be less. To view these changes, the reader is referred to the computer output part C pages C1-60.[1]

---

[1]This output is available to the reader on request
to the author.

APPENDIX D

Changes In The Discount Rate

The next factor, whose sensitivity we wanted to test in regard to investment and the objective function, that was allowed to vary was the discount rate. In changing the discount rate, not only did the upper bound on the present value of product change for every rate tested, but both the investment policy and the after-tax earnings function where as found to be sensitive to changes in the discount rate.

Considering first changes in the upper bound on the present value of product, tables D-1 - D-11 demonstrate the effect on the present value of product as the discount rate is varied from .04 to 14 percent. As is demonstrated in the tables, as the discount rate increases, the present value of product at each of the investment possibilities decreases. Thus, the higher the discount rate, the greater will be the production rate at the beginning of the life period of the reservoir. The result of the higher discount rate is to shift production to the present and thus allow

for greater production at the beginning of the well life and
in the future to produce at lower rates.

This result can be clearly seen in tables D-12,
D-13 and D-14. In table D-12, we allow the discount rate to
be .05. As can be seen from the table, the rate of production is 145 units of which 68 are from the new facilities
that were brought into solution. In table D-13 we see now
the discount rate is .09 and the number of units of production for all periods is 150 units of which 74 units are from
the new investment facilities. Thus production in every
period has been increased by 5 units overall and by 6 units
of new facilities by the increase in the discount rate. In
table D-14 the discount rate is now .14 and the production
at each facility is maximum, 75 units at the beginning of
the period in time period t=1. At t=10, the production rate
for the reservoir declines to only one facility being utilized and that one has a production rate of 33 units per
period until the end of the horizon. Thus production was
slightly shifted to the present and as the upper bound on
the present value of product declined, the utilization of
the reservoir declined.

Again, one of the limiting factors was the imposi-
tion of an upper bound on the debt limit. If this restraint

Were removed, all facilities at different rates of production would have more adequately demonstrated the effect that changes in the discount rate would have had on the shift of production over changes in the discount rate.

Table D-1. The Upper Bound on the Present Value of
Product: Discount Rate Is . 4

Time/Facility	1	2	3	4
1	1501811	1501762	1501762	1501738
2	1444049	1444002	1444002	1443978
3	1388508	1388464	1388464	1388442
4	1335104	1335061	1335061	1335040
5	1283754	1283713	1283713	1283692
6	1234379	1234339	1234339	1234320
7	1186903	1186865	1186865	1186846
8	1141253	1141216	1141216	1141198
9	1097358	1097323	1097323	1097306
10	1055152	1055118	1055118	1055102
11	1014569	1014537	1014537	1014521
12	975548	975516	975516	975501
13	938026	937996	937996	937981
14	901949	901920	901920	901905
15	867258	867230	867230	867217
16	833902	833875	833875	833862
17	801829	801803	801803	801709
18	770989	770965	770965	770952
19	741336	741312	741312	741300
20	712823	712800	712800	712789
21	685407	685385	685385	685374
22	659045	659024	659024	659013
23	633697	633677	633677	633667
24	609324	609305	609305	609296
25	585889	585870	585870	585860

Table D-2.   The Upper Bound On the Present Value of Product:
             Discount Rate Is .05

Time/Facility	1	2	3	4
1	1341947	1341899	1341923	1341923
2	1278045	1278000	1278022	1278022
3	1217186	1217142	1217164	1217164
4	1159224	1159183	1159204	1159204
5	1104023	1103984	1104004	1104004
6	1051451	1051413	1051432	1051432
7	1001382	1001346	1001364	1001364
8	953697	953663	953680	953680
9	908283	908230	908267	908267
10	865031	865000	865016	865016
11	823839	823810	823824	823829
12	784609	784581	784595	784593
13	747246	747220	747233	747233
14	711663	711638	711651	711651
15	677774	677750	677762	677762
16	645500	645477	645488	645488
17	614761	614740	614151	614751
18	585487	585466	585477	585477
19	557607	557581	557597	557597
20	531054	531035	531045	531045
21	505766	505748	505157	505757
22	481682	481665	481673	481613
23	458744	458728	458736	458736
24	436899	436889	436892	436892
25	416095	416080	416087	416087

Table D-3.   Upper Bound on the Present Value of Product:
Discount Rate Is .06

Time/Facility	1	2	3	4
1	1205670	1205623	1205623	1255600
2	1137425	1137380	1137380	1137358
3	1073042	1073000	1073000	1072979
4	1012304	1012265	1012265	1012245
5	955004	954867	954967	954948
6	900947	900912	900912	900894
7	849950	849917	849917	849900
8	801840	801808	801808	801793
9	756453	756423	756423	756408
10	713634	713607	713607	713593
11	673240	673214	673214	673201
12	635123	635107	635107	635095
13	599181	599158	599158	599146
14	565265	565243	565243	565232
15	533269	533248	533248	533238
16	503084	503064	503064	503055
17	474608	474589	474589	474580
18	447743	477126	447726	447717
19	422399	422383	422383	422374
20	398490	398474	398474	398466
21	375934	375919	375919	375912
22	354654	354641	354641	354634
23	334580	334567	334567	334560
24	315641	315629	315629	315623
25	297775	297763	297763	297757

402

Table D-4. Upper Bound on the Present Value of Product:
Discount Rate Is .07

Time/Facility	1	2	3	4
1	1088793	1088746	1088769	1088769
2	1017563	1017520	1017542	1017542
3	950994	950953	950973	950973
4	888779	888741	888760	888760
5	830635	830599	830617	830617
6	776294	776261	776278	776278
7	725509	725477	725493	725493
8	678045	678016	678031	678031
9	633687	633660	633674	633674
10	592231	592206	592218	592218
11	553487	553463	553475	553475
12	517278	517255	517267	517267
13	483437	483416	483427	483427
14	451810	451791	451801	451801
15	422253	422235	422244	422244
16	394629	394612	394620	394620
17	368812	368796	368804	368804
18	344684	344669	344677	344677
19	322135	322121	322128	322128
20	301060	301047	301054	301054
21	281365	281353	281359	281359
22	262958	262946	262952	262952
23	245755	245744	245750	245750
24	229677	229668	229673	229673
25	214652	219643	214647	214647

Table D-5. Upper Bound on the Present Value of Product:
Discount Rate is .08

Time/Facility	1	2	3	4
1	988104	988058	988058	988035
2	914911	914869	914869	914847
3	847140	847100	847100	847081
4	784389	784352	784352	784334
5	726286	726252	726252	726235
6	672487	672456	672456	672440
7	622673	622644	622644	622630
8	576549	576522	576522	576509
9	533842	533817	533817	533804
10	494298	494275	494275	494263
11	457683	457662	457662	457651
12	423781	423761	423761	423151
13	392390	392371	392371	392362
14	363324	363307	363307	363298
15	336411	336395	336395	336387
16	311492	311477	311477	311470
17	288418	288405	288405	288398
18	267054	267041	267041	267035
19	247272	247261	247261	247255
20	228956	228945	228945	228940
21	211996	211986	211986	211981
22	196293	196283	196283	196279
23	181752	181744	181744	181740
24	168289	168281	168281	168277
25	155823	155816	155816	155812

Table D-6. Upper Bound on the Present Value of Product:
Discount Rate Is .09

Time/Facility	1	2	3	4
1	900856	900810	900810	900787
2	826473	826431	826431	826410
3	758232	758194	758194	758174
4	695626	695591	695591	695573
5	638189	638157	638157	638140
6	585495	585465	585465	585450
7	537151	537124	537124	537110
8	492779	492774	492774	492761
9	452109	452086	452086	452075
10	414779	414758	414758	414747
11	380531	380512	380512	380502
12	349111	349093	349093	349085
13	320286	320269	320269	320261
14	292840	293825	293825	293818
15	269578	269564	269564	269557
16	247318	247307	247307	247300
17	226898	226887	226887	226881
18	208164	208153	208153	208148
19	190976	190966	190966	190961
20	175207	175198	175198	175194
21	160741	160732	160732	160728
22	147468	147461	147461	147457
23	135292	135285	135285	135282
24	124121	124115	124115	124112
25	113873	113867	113867	113864

Table D-7. Upper Bound on the Present Value of Product:
Discount Rate Is .10

Time/Facility	1	2	3	4
1	824867	824845	824845	824822
2	749879	749859	749859	749838
3	681708	681690	681690	681671
4	619735	619718	619718	619701
5	563395	563380	563380	563364
6	512178	512164	512164	512149
7	465616	465603	465603	465590
8	423287	423276	423276	423264
9	384807	824796	384796	384785
10	349824	349815	349815	349805
11	318022	318013	318013	318005
12	289111	289103	289103	289095
13	262828	262821	262821	262814
14	238935	238828	238828	238922
15	217213	217207	217207	217201
16	197467	197461	197461	197456
17	179515	179510	179510	179505
18	163196	163191	163191	168187
19	148360	148356	148356	148351
20	134872	134869	134869	134865
21	122611	122608	122608	122605
22	111465	111462	111462	111459
23	101332	101329	101329	101326
24	92120	92117	92117	92115
25	83745	83743	83743	83741

406

Table D-8.   Upper Bound on the Present Value of Product:
            Discount Rate Is .11

Time/Facility	1	2	3	4
1	758423	758378	758378	758355
2	683264	683223	683223	683203
3	615553	615517	615517	615498
4	554552	554519	554519	554503
5	499597	499567	499567	499552
6	450087	450060	450060	450047
7	405484	405460	405460	405448
8	365301	365279	365279	365268
9	329100	329080	329080	329070
10	296486	296469	296469	296460
11	267105	267089	267089	267081
12	240635	240621	240621	240614
13	216788	216775	216775	216769
14	195305	195293	195293	195287
15	175950	175940	175940	175935
16	158514	158504	158504	158500
17	142805	142797	142797	142792
18	128653	128646	128646	128642
19	115904	115897	115897	115894
20	104418	104412	104412	104409
21	94070	94065	94065	94062
22	84748	84743	84743	84740
23	76349	76345	76345	76343
24	68783	68779	68779	68777
25	61967	61963	61963	61961

Table D-9. Upper Bound on the Present Value of Product:
Discount Rate is .12

Time/Facility	1	2	3	4
1	699990	699945	699945	699923
2	624991	624951	624951	624931
3	558028	557992	557992	557974
4	498239	498207	498207	498191
5	444856	444828	444828	444814
6	397193	397168	397168	397155
7	354637	354614	354614	354603
8	316640	316620	316620	316610
9	282714	282696	282696	282687
10	252423	252407	252407	252399
11	225378	225364	225364	225357
12	201230	201218	201218	201211
13	179670	179659	179659	179653
14	160420	160409	160409	160404
15	143232	143223	143223	143218
16	127885	127877	127877	127873
17	114184	114176	114776	114173
18	101950	101943	101943	101940
19	91026	91021	91021	91018
20	81274	81268	81268	81266
21	72566	72561	72561	72559
22	64791	64787	64787	64785
23	57849	57845	57845	57843
24	51651	51648	51648	51646
25	46117	46114	46114	46112

Table D-10. Upper Bound on the Present Value of Product:
Discount Rate Is .13

Time/Facility	1	2	3	4
1	648384	648339	648339	648317
2	573791	573752	573752	573732
3	507779	507745	507745	507728
4	449362	449332	449332	449316
5	397666	397639	397639	397625
6	351917	351893	351893	351881
7	311431	311409	311409	311399
8	275602	275584	275584	275574
9	243896	243879	243879	243871
10	215837	215822	215822	215815
11	191006	190993	190993	190987
12	169032	169021	169021	169015
13	149586	149576	149576	149571
14	132377	132368	132368	132363
15	117148	117140	117140	117136
16	103671	103663	103663	103660
17	91744	91738	91738	91734
18	81189	81124	81184	81181
19	71849	71844	71844	71842
20	63583	63579	63579	63577
21	56268	56264	56264	56262
22	49795	49791	49791	49790
23	44066	44063	44063	44062
24	38997	38994	38994	38993
25	34510	34508	34508	34507

Table D-11. Upper Bound on the Present Value of Product:
Discount Rate Is .14

Time/Facility	1	2	3	4
1	602603	602559	602559	602537
2	528599	528561	528561	528542
3	463684	463650	463650	463633
4	406740	406710	406710	406696
5	356790	356784	356764	356751
6	312973	312950	312950	312939
7	274538	274518	274518	274518
8	240823	240805	240805	240796
9	211248	211233	211233	211225
10	185305	185292	185292	185285
11	162549	162537	162537	162531
12	142586	142576	142576	142571
13	125076	125067	125067	125062
14	109716	109708	109708	109704
15	96242	96235	96235	96231
16	84423	84416	84416	84413
17	74055	74050	74050	74047
18	64960	64956	64956	64953
19	56983	56979	56979	56977
20	49985	49981	49981	49979
21	43846	43843	43843	43842
22	38462	38459	38459	38458
23	33738	33936	33936	33735
24	29595	29593	29593	29592
25	25961	25959	25959	25958

APPENDIX E

Changes In The Interest Rate

The interest rate on savings was allowed to vary
between .05 percent and .14 percent to see the effects the
change of this variable would have on the optimal investment
policy and the objective function. Holding the other parameters of the model constant and allowing only for variations
in the interest rate indicated that interest rates had no, or
an insignificant, effect on the optimal investment policy and
the objective function. The general conclusion that can be
made is that it had a very slight effect on the debt limit
as is discernable from table E-1. The debt limit was calculated in the program as follows:

Debt Limit $=$ Initial Debt INV(N) - TEARN where
INV(N) is the investment required for the Nth facility, (N
here is 1-4) and TEARN is the total earnings derived from
the optimal operation of the N facilities.

Table E-1. Changes in the Debt Limit as Interest Rate
            Changes (In $)

Time/Interest Rate	.04	.05	.06	.07	.08
1	-1150	-1150	-1150	-1150	-1150
2	- 952	- 973	- 979	- 979	- 982
3	- 765	- 807	- 813	- 813	- 816
4	- 575	- 637	- 643	- 643	- 646
5	- 380	- 464	- 470	- 470	- 473
6	- 182	- 287	- 293	- 293	- 296
7	22	- 107	- 113	- 113	- 116
8	229	78	71	71	68
9	438	267	260	260	257
10	647	456	449	449	446
11	856	645	638	638	635
12	1065	717	710	710	707
13	1274	789	782	782	779
14	1483	861	854	854	851
15	1692	933	926	926	923
16	1901	1005	998	998	995
17	2110	1077	1070	1070	1067
18	2319	1149	1142	1142	1139
19	2528	1221	1214	1214	1211
20	2737	1293	1286	1286	1283
21	2946	1365	1358	1358	1355
22	3155	1437	1430	1430	1427
23	3664	1509	1502	1502	1499
24	3573	1581	1574	1574	1571
25	3782	1653	1646	1646	1643

.09	.10	.11	.12	.13	.14
-1150	-1150	-1150	-1150	-1150	-1150
- 985	- 988	- 990	-2917	- 996	- 997
- 819	- 822	- 824	-2792	- 830	- 831
- 649	- 652	- 654	-2665	- 660	- 661
- 476	- 479	- 481	-2534	- 487	- 488
- 299	- 302	- 304	-2400	- 310	- 311
- 119	- 112	- 124	-2264	- 130	- 131
65	62	60	-2124	54	53
254	251	249	-1982	243	242
443	440	438	-1837	432	431
632	629	627	-1688	621	620
704	701	699	-1536	693	692
776	773	771	-1382	765	764
848	845	843	-1224	837	836
920	917	915	-1062	909	908
992	989	987	- 898	981	980
1064	1061	1059	- 730	1053	1052
1136	1133	1131	- 599	1125	1124
1208	1205	1203	- 384	1197	1196
1280	1277	1275	- 205	1269	1268
1352	1349	1347	- 23	1341	1340
1424	1421	1419	164	1413	1412
1496	1493	1491	353	1485	1484
1568	1565	1563	542	1557	1556
1640	1637	1635	614	1629	1628

APPENDIX F

Changes In Debt/Equity Ratio

Changes were also allowed in the relation between debt and equity to see what effects changes in this variable would have on the objective function, and ultimately on the optimal investment policy.

The debt/equity ratio was allowed to vary from .01 to .40. The results indicated that the only variable which changed over time and over the changes in the debt/equity ratio was the allowable debt limit. As the debt/equity ratio increased, not surprisingly, the allowable debt limit increased. However, this increase in the debt limit did not result in a change in the objective function and likewise did not result in new optimal patterns of investment.

The general conclusion then is that altering the debt/equity ratio will not alter the optimal investment patterns if unaccompanied by changes in debt itself. The only change will be in the allowable debt limit which will have no effect on the optimal policy.

Table F-1. Changes in the Debt/Equity Ratio

Time/ETA	.01		.02	
	Debt	Debt Limit	Debt	Debt Limit
1	1250	-1240	1250	-1230
2	1082	-1072	1063	-1043
3	916	- 906	877	- 857
4	746	- 736	687	- 667
5	573	- 563	493	- 473
6	396	- 386	295	- 275
7	216	- 206	92	- 72
8	32	- 22	-115	135
9	-157	167	-324	344
10	-346	356	-533	553
11	-535	545	-742	762
12	-607	617	-951	971
13	-679	689	-1160	1180
14	-751	761	-1369	1389
15	-823	833	-1578	1598
16	-895	905	-1787	1807
17	-967	977	-1996	2016
18	-1039	1049	-2205	2225
19	-1111	1121	-2414	2434
20	-1183	1193	-2623	2643
21	-1255	1265	-2832	2852
22	-1327	1337	-3041	3061
23	-1399	1409	-3250	3270
24	-1471	1481	-3459	3479
25	-1543	1553	-3668	3688

Time/ETA	.03		.04	
	Debt	Debt Limit	Debt	Debt Limit
1	1250	-1220	1250	-1210
2	1082	-1052	1082	-1042
3	916	- 976	916	- 976
4	746	- 716	746	- 706
5	573	- 543	573	- 533
6	396	- 366	396	- 356
7	216	- 186	216	- 176
8	32	- 2	32	8
9	-157	187	-157	197
10	-346	376	-346	386
11	-535	565	-535	575
12	-607	637	-607	647
13	-679	709	-679	719
14	-751	781	-751	791
15	-823	853	-823	863
16	-895	925	-895	935
17	-967	997	-967	1007
18	-1039	1069	-1039	1079
19	-1111	1141	-1111	1151
20	-1183	1213	-1183	1223
21	-1255	1285	-1255	1295
22	-1327	1357	-1327	1367
23	-1399	1429	-1399	1439
24	1471	1501	-1471	1511
25	1543	1573	-1543	1583

Time/ETA	.05		.06	
	Debt	Debt Limit	Debt	Debt Limit
1	1250	-1200	1250	-1190
2	1082	-1032	1082	-1022
3	916	-866	916	-956
4	746	-696	746	-686
5	573	-523	573	-513
6	396	-346	396	-336
7	216	-166	216	-156
8	32	18	32	28
9	-157	207	-157	217
10	-346	396	-346	406
11	-535	585	-535	595
12	-607	657	-607	667
13	-679	729	-697	739
14	-751	801	-751	811
15	-823	873	-823	883
16	-895	945	-895	955
17	-967	1017	-967	1027
18	-1039	1089	-1039	1099
19	-1111	1161	-1111	1171
20	-1183	1233	-1183	1243
21	-1255	1305	-1255	1315
22	-1327	1377	-1327	1387
23	-1399	1449	-1399	1459
24	-1471	1521	-1471	1531
25	-1543	1593	-1543	1603

Time/ETA	.07		.08	
	Debt	Debt Limit	Debt	Debt Limit
1	1250	-1180	1250	-1170
2	1082	-1012	1082	-1002
3	916	-846	916	-836
4	746	-676	746	-666
5	573	-503	573	-493
6	396	-326	396	-316
7	216	-146	216	-136
8	32	38	32	48
9	-157	227	-157	237
10	-346	416	-346	426
11	-535	605	-535	615
12	-607	677	-607	687
13	-679	749	-679	759
14	-151	821	-751	831
15	-823	893	-823	903
16	-895	965	-895	975
17	-967	1037	-967	1047
18	-1039	1109	-1039	1119
19	-1111	1181	-1111	1191
20	-1183	1253	-1183	1263
21	-1255	1325	-1255	1335
22	-1327	1397	-1327	1479
23	-1399	1469	-1397	
24	-1471	1541	-1471	1551
25	-1543	1613	-1543	1623

Time/ETA	.09		.10	
	Debt	Debt Limit	Debt	Debt Limit
1	1250	-1160	1250	-1150
2	1063	-973	1079	-979
3	877	-787	916	-816
4	687	-597	746	-646
5	493	-403	573	-473
6	295	-205	396	-296
7	92	-2	216	-116
8	-115	205	32	68
9	-324	414	-157	257
10	-533	623	-346	446
11	-742	832	-535	635
12	-951	1041	-607	707
13	-1160	1250	-679	779
14	-1369	1459	-751	851
15	-1578	1668	-823	923
16	-1787	1877	-895	995
17	-1996	2086	-967	1067
18	-2205	2295	-1039	1139
19	-2414	2504	-1111	1211
20	-2623	2713	-1183	1283
21	-2832	2922	-1225	1325
22	-3041	3131	-1327	1427
23	-3250	3340	-1397	1497
24	-3459	3549	-1491	1491
25	-3668	3758	-1543	1643

Time/ETA	.11		.12	
	Debt	Debt Limit	Debt	Debt Limit
1	1250	-1140	1250	-1130
2	1082	-972	1082	-962
3	916	-806	916	-796
4	746	-636	746	-626
5	573	-463	573	-453
6	396	-286	396	-276
7	216	-106	216	-96
8	32	78	32	88
9	-157	267	157	277
10	-346	456	-346	466
11	-535	645	-535	655
12	-607	717	-607	727
13	-679	789	-679	799
14	-751	861	-751	871
15	-823	933	-823	943
16	-895	1005	-895	1015
17	-967	1077	-967	1087
18	-1039	1149	-1037	1159
19	-1111	1221	-1111	1231
20	-1183	1293	-1183	1303
21	-1255	1365	-1255	1375
22	-1327	1437	-1327	1447
23	-1399	1509	-1399	1519
24	-1471	1591	-1477	1591
25	-1543	1653	-1543	1663

Time/ETA	.13		.14	
	Debt	Debt Limit	Debt	Debt Limit
1	1250	-1120	1250	-1110
2	1082	-952	1082	-942
3	916	-786	916	-776
4	746	-616	746	-606
5	573	-443	573	-433
6	396	-226	396	-216
7	216	-86	216	-76
8	32	98	32	108
9	-157	287	-157	297
10	-346	476	-346	486
11	-535	665	-535	675
12	-607	737	-607	747
13	-679	809	-679	819
14	-751	881	-751	891
15	-823	953	-823	963
16	-895	1025	-895	1035
17	-967	1079	-967	1107
18	-1039	1169	-1039	1179
19	-1111	1241	-1111	1351
20	-1183	1313	-1183	1323
21	-1255	1385	-1255	1395
22	-1327	1457	-1327	1467
23	-1399	1529	-1399	1539
24	-1471	1601	-1471	1611
25	-1543	1673	-1543	1683

Time/ETA	.15 Debt	Debt Limit	.16 Debt	Debt Limit
1	1250	-1100	1250	-1090
2	1082	-032	1082	-922
3	916	-766	916	-756
4	746	-596	746	-586
5	573	-423	573	-413
6	396	-206	396	-196
7	216	-66	216	-56
8	32	118	32	128
9	-157	307	-157	317
10	-346	496	-346	506
11	-535	685	-535	695
12	-607	757	-607	767
13	-679	829	-679	839
14	-751	901	-751	911
15	-832	973	-823	983
16	-895	1045	-895	1055
17	-967	1117	-967	1127
18	-1039	1189	-1039	1199
19	-1111	1361	-1111	1321
20	-1183	1333	-1183	1343
21	-1255	1405	-1255	1415
22	-1327	1477	-1327	1487
23	-1399	1549	-1399	1559
24	-1471	1621	-1471	1631
25	-1543	1693	-1543	1703

Time/ETA	.17		.18	
	Debt	Debt Limit	Debt	Debt Limit
1	1250	-1080	1250	-1070
2	1063	-893	1082	-902
3	877	-707	916	-736
4	687	-517	746	-566
5	493	-323	573	-393
6	295	-125	396	-216
7	92	78	216	-36
8	-115	285	32	148
9	-324	494	-157	337
10	-533	703	-346	526
11	-742	912	-418	598
12	-951	1121	-490	670
13	-1160	1330	-562	742
14	-1369	1539	-634	814
15	-1578	1748	-706	886
16	-1787	1957	-778	958
17	-1996	2166	-850	1030
18	-2205	2375	-922	1102
19	-2414	2584	-994	1174
20	-2623	2793	-1066	1246
21	-2832	3002	-1138	1318
22	-3041	3211	-1210	1390
23	-3250	3420	-1282	1462
24	-3459	3629	-1359	1534
25	-3668	3838	-1426	1606

Time/ETA	.19 Debt	Debt Limit	.20 Debt	Debt Limit
1	1250	-1060	1250	-1050
2	1082	-892	1082	-882
3	916	-726	916	-716
4	746	-556	746	-546
5	573	-383	573	-373
6	396	-206	396	-196
7	216	-26	216	-16
8	32	158	32	168
9	-157	347	-157	357
10	-346	536	-346	546
11	-535	725	-418	618
12	-607	797	-490	690
13	-697	869	-562	762
14	-751	941	-634	834
15	-832	1013	-706	906
16	-895	1085	-778	978
17	-967	1157	-850	1050
18	-1039	1229	-992	1122
19	-1111	1301	-994	1194
20	-1183	1373	-1066	1266
21	-1255	1445	-1138	1338
22	-1327	1517	-1210	1410
23	-1399	1589	-1282	1482
24	-1471	1661	-1354	1554
25	-1543	1733	-1426	1626

Time/ETA	.21 Debt	Debt Limit	.22 Debt	Debt Limit
1	1250	-1040	1250	-1030
2	1082	-872	1082	-862
3	916	-706	916	-696
4	746	-536	746	-526
5	573	-363	573	-358
6	396	-186	396	-176
7	216	-6	216	4
8	32	178	32	188
9	-157	367	-157	377
10	-346	556	-346	566
11	-418	628	-418	638
12	-490	700	-490	710
13	-562	772	-562	782
14	-634	844	-634	854
15	-706	916	-706	926
16	-778	988	-778	998
17	-850	1060	-850	1070
18	-922	1132	-922	1142
19	-994	1204	-994	1214
20	-1066	1276	-1066	1286
21	-1138	1348	-1138	1358
22	-1210	1420	-1210	1430
23	-1282	1492	-1282	1502
24	-1359	1564	-1354	1574
25	-1426	1636	-1426	1646

Time/ETA	.23 Debt	Debt Limit	.24 Debt	Debt Limit
1	1250	-1020	1250	-1010
2	1082	-852	1063	-823
3	916	-686	877	-637
4	746	-516	687	-447
5	573	-343	493	-253
6	396	-166	295	-55
7	216	14	92	148
8	32	198	-115	355
9	-157	387	-324	564
10	-346	576	-533	773
11	-418	648	-742	982
12	-490	720	-951	1191
13	-562	792	-1160	1400
14	-634	864	-1369	1609
15	-706	936	-1578	1818
16	-778	1008	-1787	2027
17	-850	1080	-1996	2236
18	-922	1152	-2205	2445
19	-994	1224	-2414	2654
20	-1066	1296	-2623	2863
21	-1138	1368	-2832	3072
22	-1210	1440	-3041	3281
23	-1282	1512	-3250	3490
24	-1354	1584	-3459	3699
25	-1426	1656	-3668	3908

Time/ETA	.25		.26	
	Debt	Debt Limit	Debt	Debt Limit
1	1250	-1000	1250	-990
2	1082	-832	1082	-822
3	916	-666	916	-656
4	746	-496	746	-486
5	573	-323	573	-313
6	396	-146	396	-136
7	214	34	216	44
8	32	218	32	228
9	-157	407	-157	417
10	-346	596	-346	606
11	-418	668	-418	678
12	-490	740	-490	750
13	-562	812	-562	822
14	-634	884	-634	894
15	-706	956	-706	966
16	-778	1028	-778	1038
17	-850	1100	-850	1110
18	-922	1172	-922	1182
19	-994	1244	-994	1254
20	1066	1316	-1066	1326
21	-1138	1388	-1138	1398
22	-1210	1460	-1210	1470
23	-1282	1532	-1282	1542
24	-1354	1604	-1354	1614
25	-1426	1676	-1426	1686

Time/ETA	.27		.28	
	Debt	Debt Limit	Debt	Debt Limit
1	1250	-980	1250	-970
2	1082	-812	1082	-802
3	916	-646	916	-636
4	746	-476	746	-466
5	573	-303	573	-293
6	396	-126	396	-116
7	216	54	216	64
8	32	238	32	248
9	-157	427	-157	437
10	-346	616	-346	626
11	-418	688	-418	698
12	-490	760	-490	770
13	-562	832	-562	842
14	-634	904	-634	914
15	-706	976	-706	986
16	-778	1048	-778	1058
17	-850	1120	-850	1130
18	-922	1192	-922	1202
19	-994	1264	-994	1274
20	-1066	1336	-1066	1346
21	-1138	1408	-1138	1418
22	-1210	1480	-1210	1490
23	-1282	1552	-1282	1562
24	-1354	1624	-1354	1634
25	-1426	1696	-1426	1706

Time/ETA	.29		.30	
	Debt	Debt Limit	Debt	Debt Limit
1	1250	-960	1250	-950
2	1082	-792	1082	-782
3	916	-626	916	-616
4	746	-456	746	-446
5	573	-283	573	-273
6	396	-106	396	-96
7	216	74	216	84
8	32	258	32	268
9	-157	447	-157	457
10	-346	636	-346	646
11	-418	708	-416	718
12	-490	780	-490	790
13	-562	852	-562	862
14	-634	924	-634	934
15	-706	996	-706	1006
16	-778	1068	-778	1078
17	-850	1140	-850	1150
18	-922	1212	-922	1222
19	-994	1284	-994	1294
20	-1066	1356	-1066	1366
21	-1138	1428	-1138	1438
22	-1210	1500	-1210	1510
23	-1282	1572	-1282	1582
24	-1354	1644	-1354	1654
25	-1426	1716	-1426	1726

Time/ETA	.31		.32	
	Debt	Debt Limit	Debt	Debt Limit
1	1250	-940	1250	-930
2	1082	-772	1082	-762
3	916	-606	914	-596
4	746	-436	746	-426
5	573	-263	573	-253
6	396	-86	396	-76
7	216	94	216	104
8	32	278	32	288
9	-157	467	-157	477
10	-346	656	-346	666
11	-418	728	-418	738
12	-490	800	-490	810
13	-562	872	-562	882
14	-634	944	-634	954
15	-706	1016	-706	1026
16	-778	1088	-778	1098
17	-850	1160	-850	1170
18	-922	1232	-922	1242
19	-994	1304	-994	1314
20	-1066	1376	-1066	1386
21	-1138	1448	-1138	1458
22	-1210	1520	-1210	1530
23	-1282	1592	-1282	1602
24	-1354	1664	-1354	1674
25	-1426	1736	-1426	1746

Time/ETA	.33		.34	
	Debt	Debt Limit	Debt	Debt Limit
1	1250	-920	1250	-910
2	1082	-752	1082	-742
3	914	-586	914	-576
4	746	-416	746	-406
5	573	-243	573	-233
6	396	-66	396	-56
7	216	114	216	124
8	32	298	32	308
9	-157	487	-157	497
10	-346	676	-346	686
11	-418	748	-418	758
12	-490	820	-490	830
13	-562	892	-562	902
14	-634	964	-634	974
15	-706	1036	-706	1046
16	-778	1108	-778	1118
17	-850	1180	-850	1190
18	-922	1252	-922	1262
19	-994	1324	-994	1334
20	-1066	1396	-1066	1406
21	-1138	1468	-1138	1478
22	-1210	1540	-1210	1550
23	-1282	1612	-1282	1622
24	-1354	1684	-1354	1694
25	-1426	1756	-1426	1766

Time/ETA	Debt	.35 Debt Limit	Debt	.36 Debt Limit
1	1250	-900	1250	-890
2	1082	-732	1082	-722
3	914	-566	914	-556
4	746	-396	746	-386
5	573	-223	573	-213
6	396	-46	396	-36
7	216	134	216	144
8	32	318	32	328
9	-157	507	-157	517
10	-346	696	-346	706
11	-418	768	-418	778
12	-490	840	-490	850
13	-562	912	-562	922
14	-634	984	-634	994
15	-706	1056	-706	1066
16	-778	1128	-778	1138
17	-850	1200	-850	1210
18	-922	1272	-922	1282
19	-994	1344	-994	1354
20	-1066	1416	-1066	1426
21	-1138	1488	-1138	1498
22	-1210	1560	-1210	1570
23	-1282	1632	-1282	1642
24	-1354	1704	-1354	1714
25	-1426	1776	-1426	1786

Time/ETA	.37		.38	
	Debt	Debt Limit	Debt	Debt Limit
1	1250	-880	1250	-870
2	1082	-712	1063	-683
3	916	-546	877	-497
4	746	-376	687	-307
5	573	-203	493	-113
6	396	-26	295	85
7	216	154	92	288
8	32	338	-115	495
9	-157	527	-324	704
10	-229	599	-533	913
11	-301	671	-742	1122
12	-373	743	-951	1331
13	-445	815	-1160	1540
14	-517	887	-1369	1749
15	-589	959	-1578	1958
16	-661	1031	-1787	2167
17	-733	1103	-1996	2376
18	-805	1175	-2205	2585
19	-877	1247	-2414	2794
20	-949	1319	-2623	3003
21	-1021	1391	-2832	3212
22	-1093	1463	-3041	3421
23	-1165	1535	-3250	3630
24	-1237	1607	-3459	3839
25	-1309	1679	-3669	4048

Time/ETA	.39		.40	
	Debt	Debt Limit	Debt	Debt Limit
1	1250	-860	1250	-850
2	1082	-692	1082	-682
3	916	-526	916	-516
4	746	-356	746	-346
5	573	-183	573	-173
6	396	-6	396	4
7	216	174	216	184
8	32	358	32	368
9	-157	547	-154	557
10	-229	619	-229	629
11	-301	691	-301	701
12	-373	763	-373	773
13	-445	835	-445	845
14	-517	907	-517	917
15	-589	979	-589	989
16	-661	1051	-661	1061
17	-773	1123	-733	1133
18	-805	1195	-805	1205
19	-877	1267	-877	1277
20	-949	1339	-949	1349
21	-1021	1411	-1021	1421
22	-1093	1483	-1093	1493
23	-1165	1555	-1165	1565
24	-1237	1627	-1237	1637
25	-1309	1699	-1309	1709

APPENDIX G

Changes In The Depletion Allowance

Another parameter of particular interest in the
oil and gas industry is the depletion allowance, and thus
our dynamic model included consideration of this factor.
In the capital investment program, the depletion factor was
allowed to vary from .00 percent to the present .275 percent.

In the program the depletion allowance was calculated in the following manner: if net income from recovery
was negative or zero, then the depletion allowance for that
facility was zero. Secondly, if there was positive net
income from recovery but production was less than XB, then
the depletion allowance was given by

367              $DA = .50$ (net income).

If net income exceeded the percent of sales for that period,
then the depletion allowance for that facility was given by

368          $DA = D1$. market price . sales

where D1 was allowed to vary between the limits zero and
27 1/2 percent. The factors of interest to us in this

appendix are: (1) the depletion allowance and how it varies
over the various rates, and (2) the effects on taxable
income, (3) and on taxes paid, and finally (4) the effects
on earnings for each well and the effects on total earnings
for the reservoir. Tables G-1 - G-28 describe the effects
on the factors mentioned above.

The general conclusions derivable from the limited
analysis performed, and it is strongly suggested that more
analysis be performed with this model before definitive
conclusions be derived, are first, that depletion allowance
does have the effect of accelerating investment in the reservoir, in this case the hypothetical case presented.
Secondly, it significantly affects the earnings of (1) the
individual well in question and (2) the total earnings for
the reservoir. To demonstrate the effects on after-tax
earnings we have given Table G-29 which gives the present
value of after-tax earnings for each depletion allowance
allowed for.

Table G-1. The Changes in Taxable Income, Taxes Paid, and
Earnings When the Depletion Allowance is .00

Facility Time	DA(I) 1	2	TI(I) 1	2	Tax (I) 1	2	Earn(I) 1	2
1	0	0	121	129	60	64	61	65
2	0	0	118	126	59	63	59	63
3	0	0	120	128	60	64	60	64
4	0	0	122	130	61	65	61	65
5	0	0	124	132	62	66	62	66
6	0	0	126	135	63	67	63	68
7	0	0	128	137	64	68	64	69
8	0	0	130	139	65	69	65	70
9	0	0	133	142	66	71	67	71
10	0	0	135	144	67	72	68	72
11	0	0	137	147	68	73	69	74
12	0	0	140	149	70	74	70	75
13	0	0	142	152	71	76	71	76
14	0	0	144	153	72	76	72	77
15	0	0	144	153	72	76	72	77
16	0	0	144	153	72	76	72	77
17	0	0	142	-20	71	-10	71	-10
18	0	0	142	-20	71	-10	71	-10
19	0	0	142	-20	71	-10	71	-10
20	0	0	142	-20	71	-10	71	-10
21	0	0	142	-20	71	-10	71	-10
22	0	0	142	-20	71	-10	71	-10
23	0	0	142	-20	71	-10	71	-10
24	0	0	142	-20	71	-10	71	-10
25	0	0	142	-20	71	-10	71	-10

Table G-2. The Changes in Taxable Income, Taxes Paid, and
Earnings When the Depletion Allowance is .01

Facility	DA(I)		TI(I)		Tax(I)		Earn(I)	
Time	1	2	1	2	1	2	1	2
1	3	3	145	127	72	63	76	67
2	3	3	141	124	70	62	74	65
3	3	3	144	127	72	63	75	67
4	3	3	147	129	73	64	77	68
5	3	3	150	132	75	66	78	69
6	3	3	153	134	76	67	80	70
7	3	3	156	137	78	68	81	72
8	3	3	159	140	79	70	83	73
9	3	3	162	143	81	71	84	75
10	3	3	165	145	82	72	86	76
11	3	3	168	148	84	74	87	77
12	3	3	171	151	85	75	89	79
13	3	3	171	151	85	75	89	79
14	3	3	171	151	85	75	89	79
15	3	3	171	151	85	75	89	79
16	3	3	171	151	85	75	89	79
17	3	0	171	-21	85	-10	89	-11
18	3	0	171	-21	85	-10	89	-11
19	3	0	171	-21	85	-10	89	-11
20	3	0	171	-21	85	-10	89	-11
21	3	0	171	-21	85	-10	89	-11
22	3	0	171	-21	85	-10	89	-11
23	3	0	171	-21	85	-10	89	-11
24	3	0	171	-21	85	-10	89	-11
25	3	0	171	-21	85	-10	89	-11

Table G-3. The Changes in Taxable Income, Taxes Paid, and
            Earnings When the Depletion Allowance is .02

Facility Time	DA(I) 1	DA(I) 2	TI(I) 1	TI(I) 2	Tax(I) 1	Tax(I) 2	Earn(I) 1	Earn(I) 2
1	5	6	116	123	58	61	63	68
2	5	6	113	120	56	60	62	66
3	5	6	115	122	57	61	63	67
4	5	6	117	124	58	62	64	68
5	5	6	119	127	59	63	65	70
6	5	6	122	129	61	64	66	71
7	5	6	124	132	62	66	67	72
8	5	6	126	134	63	67	68	73
9	5	6	129	137	64	68	70	75
10	5	6	131	140	65	70	71	76
11	5	6	134	142	67	71	72	77
12	5	6	137	145	68	72	74	79
13	5	6	139	147	69	73	75	80
14	5	6	139	147	69	73	75	80
15	5	6	139	147	69	73	75	80
16	5	6	139	147	69	73	75	80
17	5	0	137	-20	68	-10	74	-1-
18	5	0	137	-20	68	-10	74	-10
19	5	0	137	-20	68	-10	74	-10
20	5	0	137	-20	68	-10	74	-10
21	5	0	137	-20	68	-10	74	-10
22	5	0	137	-20	68	-10	74	-10
23	5	0	137	-20	68	-10	74	-10
24	5	0	137	-20	68	-10	74	-10
25	5	0	137	-20	68	-10	74	-10

Table G-4. The Changes in Taxable Income, Taxes Paid, and
Earnings When the Depletion Allowance is .03

Facility Time	DA(I) 1	DA(I) 2	TI(I) 1	TI(I) 2	Tax(I) 1	Tax(I) 2	Earn (I) 1	Earn (I) 2
1	8	9	113	120	56	60	65	69
2	8	9	110	117	55	58	63	68
3	8	9	112	119	56	59	64	69
4	8	9	114	122	57	61	65	70
5	8	9	116	124	58	62	66	71
6	8	9	119	127	59	63	68	73
7	8	9	121	129	60	64	69	74
8	8	9	124	132	62	66	70	75
9	8	9	126	134	63	67	71	76
10	8	9	129	137	64	68	73	78
11	8	9	131	140	65	70	74	79
12	8	9	134	143	67	71	75	81
13	8	9	136	144	68	72	76	81
14	8	9	136	144	68	72	76	81
15	8	9	136	144	68	72	76	81
16	8	0	134	-20	67	-10	75	-10
17	8	0	134	-20	67	-10	75	-10
18	8	0	134	-20	67	-10	75	-10
19	8	0	134	-20	67	-10	75	-10
20	8	0	134	-20	67	-10	75	-10
21	8	0	134	-20	67	-10	75	-10
22	8	0	134	-20	67	-10	75	-10
23	8	0	134	-20	67	-10	75	-10
24	8	0	134	-20	67	-10	75	-10
25	8	0	134	-20	67	-10	75	-10

Table G-5. The Changes in Taxable Income, Taxes Paid, and
Earnings When the Depletion Allowance is .03

Facility Time	DA(I) 1	2	TI(I) 1	2	Tax(I) 1	2	Earn (I) 1	2
1	11	12	104	111	52	55	63	68
2	11	12	101	108	50	54	62	66
3	11	12	103	110	51	55	63	67
4	11	12	105	112	52	56	64	68
5	11	12	108	115	54	57	65	70
6	11	12	110	117	55	58	66	71
7	11	12	112	120	56	60	67	72
8	11	12	115	122	57	61	69	73
9	11	12	117	125	58	62	70	75
10	11	12	120	127	60	63	71	76
11	11	12	122	130	61	65	72	77
12	11	12	125	133	62	66	74	79
13	11	12	127	135	63	67	75	80
14	11	12	127	135	63	67	75	80
15	11	12	127	135	63	67	75	80
16	11	12	127	135	63	67	75	80
17	11	0	124	-20	62	-10	73	-10
18	11	0	124	-20	62	-10	73	-10
19	11	0	124	-20	62	-10	73	-10
20	11	0	124	-20	62	-10	73	-10
21	11	0	124	-20	62	-10	73	-10
22	11	0	124	-20	62	-10	73	-10
23	11	0	124	-20	62	-10	73	-10
24	11	0	124	-20	62	-10	73	-10
25	11	0	124	-20	62	-10	73	-10

Table G-6. The Changes in Taxable Income, Taxes Paid, and
Earnings When the Depletion Allowance is .05

Facility Time	DA(I) 1	2	TI(I) 1	2	Tax(I) 1	2	Earn(I) 1	2
1	14	15	96	102	48	51	62	66
2	14	15	92	99	46	49	60	65
3	14	15	94	101	47	50	61	66
4	14	15	97	103	48	51	63	67
5	14	15	99	105	49	52	64	68
6	14	15	101	108	50	54	65	69
7	14	15	103	110	51	55	66	70
8	14	15	105	113	52	56	67	72
9	14	15	108	115	54	57	68	73
10	14	15	110	118	55	59	69	74
11	14	15	113	120	56	60	71	75
12	14	15	115	123	57	61	72	77
13	14	15	118	126	59	63	73	78
14	14	15	118	126	59	63	73	78
15	14	15	118	126	59	63	73	78
16	14	15	118	126	59	63	73	78
17	14	0	113	-20	56	-10	71	-10
18	14	0	113	-20	56	-10	71	-10
19	14	0	113	-20	56	-10	71	-10
20	14	0	113	-20	56	-10	71	-10
21	14	0	113	-20	56	-10	71	-10
22	14	0	113	-20	56	-10	71	-10
23	14	0	113	-20	56	-10	71	-10
24	14	0	113	-20	56	-10	71	-10
25	14	0	113	-20	56	-10	71	-10

Table G-7. The Changes in Taxable Income, Taxes Paid, and
Earnings When the Depletion Allowance is .06

Facility Time	DA(I) 1	2	TI(I) 1	2	Tax(I) 1	2	Earn(I) 1	2
1	17	18	87	93	43	46	61	65
2	17	18	84	90	42	45	59	63
3	17	18	86	92	43	46	60	64
4	17	18	88	94	44	47	61	65
5	17	18	90	96	45	48	62	66
6	17	18	92	99	46	49	63	68
7	17	18	94	101	47	50	64	69
8	17	18	96	103	48	51	65	70
9	17	18	99	106	49	53	67	71
10	17	18	101	108	50	54	68	72
11	17	18	103	111	51	55	69	74
12	17	18	106	113	53	56	70	75
13	17	18	108	116	54	58	71	76
14	17	18	110	117	55	58	72	77
15	17	18	110	117	55	58	72	77
16	17	18	110	111	55	58	72	77
17	17	0	103	-20	51	-10	69	-10
18	17	0	103	-20	51	-10	69	-10
19	17	0	103	-20	51	-10	69	-10
20	17	0	103	-20	51	-10	69	-10
21	17	0	103	-20	51	-10	69	-10
22	17	0	103	-20	51	-10	69	-10
23	17	0	103	-20	51	-10	69	-10
24	17	0	103	-20	51	-10	69	-10
25	17	0	103	-20	51	-10	69	-10

Table G-8. The Changes in Taxable Income, Taxes Paid, and
Earnings When the Depletion Allowance is .07

Facility Time	DA(I) 1	2	TI(I) 1	2	Tax(I) 1	2	Earn(I) 1	2
1	20	22	101	107	50	53	71	76
2	20	22	908	104	49	52	69	74
3	20	22	100	107	50	53	70	76
4	20	22	103	110	51	55	72	77
5	20	22	106	112	53	56	73	78
6	20	22	108	115	54	57	74	80
7	20	22	111	118	55	59	76	81
8	20	22	114	121	57	60	77	83
9	20	22	117	124	52	62	79	84
10	20	22	119	127	59	63	80	86
11	20	22	122	130	61	65	81	87
12	20	22	124	131	62	65	82	88
13	20	22	124	131	62	65	82	88
14	20	22	124	131	62	65	82	88
15	20	22	124	131	62	65	82	88
16	20	22	124	131	62	65	82	88
17	20	0	122	-20	61	-10	81	-10
18	20	0	122	-20	61	-10	81	-10
19	20	0	122	-20	61	-10	81	-10
20	20	0	122	-20	61	-10	81	-10
21	20	0	122	-20	61	-10	81	-10
22	20	0	122	-20	61	-10	81	-10
23	20	0	122	-20	61	-10	81	-10
24	20	0	122	-20	61	-10	81	-10
25	20	0	122	-20	61	-10	81	-10

Table G-9. The Changes in Taxable Income, Taxes Paid, and
Earnings When the Depletion Allowance is .08

Facility Time	DA(I) 1	2	TI(I) 1	2	Tax(I) 1	2	Earn(I) 1	2
1	28	25	120	105	60	52	88	78
2	28	25	117	103	58	51	81	77
3	28	25	120	106	60	53	88	78
4	28	25	124	109	62	54	90	80
5	28	25	127	112	63	56	92	81
6	28	25	131	115	65	57	94	83
7	28	25	134	118	67	59	95	84
8	28	25	138	121	69	60	97	86
9	28	25	142	125	71	62	99	88
10	28	25	145	128	72	64	101	89
11	28	25	146	129	73	64	101	90
12	28	25	146	129	73	64	101	90
13	28	25	146	129	73	64	101	90
14	28	25	146	129	73	64	101	90
15	28	25	146	129	73	64	101	90
16	28	25	146	129	73	64	101	90
17	28	0	140	-20	71	-10	98	-10
18	28	0	140	-20	71	-10	98	-10
19	28	0	140	-20	71	-10	98	-10
20	28	0	140	-20	71	-10	98	-10
22	28	0	140	-20	71	-10	98	-10
23	28	0	140	-20	71	-10	98	-10
24	28	0	140	-20	71	-10	98	-10
25	28	0	140	-20	71	-10	98	-10

Table G-10. The Changes in Taxable Income, Taxes Paid, and
Earnings When the Depletion Allowance is .01

	DA(I)		TI(I)		Tax(I)		Earn(I)	
Facility	1	2	1	2	1	2	1	2
Time								
1	26	28	95	101	47	50	74	79
2	26	28	92	98	46	49	72	77
3	26	28	95	101	47	50	74	79
4	26	28	97	104	48	52	75	80
5	26	28	100	107	50	53	76	82
6	26	28	103	110	51	55	78	83
7	26	28	106	113	53	56	79	85
8	26	28	109	116	54	58	81	86
9	26	28	112	119	56	59	82	88
10	26	28	115	122	57	61	84	89
11	26	28	118	125	59	62	85	91
12	26	28	118	125	59	62	85	91
13	26	28	118	125	59	62	85	91
14	26	0	116	-20	58	-10	84	-10
15	26	0	116	-20	58	-10	84	-10
16	26	0	116	-20	58	-10	84	-10
17	26	0	116	-20	58	-10	84	-10
18	26	0	116	-20	58	-10	84	-10
19	26	0	116	-20	58	-10	84	-10
20	26	0	116	-20	58	-10	84	-10
21	26	0	116	-20	58	-10	84	-10
22	26	0	116	-20	58	-10	84	-10
23	26	0	116	-20	58	-10	84	-10
24	26	0	116	-20	58	-10	84	-10
25	26	0	116	-20	58	-10	84	-10

Table G-11. The Changes in Taxable Income, Taxes Paid, and
Earnings When the Depletion Allowance is .10

Facility Time	DA(I) 1	2	TI(I) 1	2	Tax(I) 1	2	Earn(I) 1	2
1	29	31	92	98	46	49	75	80
2	29	31	89	95	44	47	74	79
3	29	31	92	98	46	49	75	80
4	29	31	95	101	47	50	77	82
5	29	31	97	104	48	52	78	83
6	29	31	100	107	50	53	79	85
7	29	31	103	110	51	55	81	86
8	29	31	106	113	53	56	82	88
9	29	31	109	116	54	58	84	89
10	29	31	112	120	56	60	85	91
11	29	31	115	122	57	61	87	92
12	29	31	115	122	57	61	87	92
13	29	31	115	122	57	61	87	92
14	29	0	113	-20	56	-10	86	-10
15	29	0	113	-20	56	-10	86	-10
16	29	0	113	-20	56	-10	86	-10
17	29	0	113	-20	56	-10	86	-10
18	29	0	113	-20	56	-10	86	-10
19	29	0	113	-20	56	-10	86	-10
20	29	0	113	-20	56	-10	86	-10
21	29	0	113	-20	56	-10	86	-10
22	29	0	113	-20	56	-10	86	-10
23	29	0	113	-20	56	-10	86	-10
24	29	0	113	-20	56	-10	86	-10
25	29	0	113	-20	56	-10	86	-10

Table G-12. The Changes in Taxable Income, Taxes Paid, and Earnings When the Depletion Allowance is .11

Facility Time	DA(I) 1	DA(I) 2	TI(I) 1	TI(I) 2	Tax(I) 1	Tax(I) 2	Earn(I) 1	Earn(I) 2
1	32	34	83	89	41	44	74	79
2	32	34	80	86	40	43	72	77
3	32	34	83	89	41	44	74	79
4	32	34	86	92	43	46	75	80
5	32	34	88	95	44	47	76	82
6	32	34	91	98	45	49	78	83
7	32	34	94	101	47	50	79	85
8	32	34	97	104	48	52	81	86
9	32	34	100	107	50	53	82	88
10	32	34	103	110	51	55	84	89
11	32	34	106	113	53	56	85	91
12	32	34	106	113	53	56	85	91
13	32	34	106	113	53	56	85	91
14	32	0	103	-20	51	-10	84	-10
15	32	0	103	-20	51	-10	84	-10
16	32	0	103	-20	51	-10	84	-10
17	32	0	103	-20	51	-10	84	-10
18	32	0	103	-20	51	-10	84	-10
19	32	0	103	-20	51	-10	84	-10
20	32	0	103	-20	51	-10	84	-10
21	32	0	103	-20	51	-10	84	-10
22	32	0	103	-20	51	-10	84	-10
23	32	0	103	-20	51	-10	84	-10
24	32	0	103	-20	51	-10	84	-10
25	32	0	103	-20	51	-10	84	-10

Table G-13. The Changes in Taxable Income, Taxes Paid, and
Earnings When the Depletion Allowance is .12

Facility Time	DA(I) 1	2	TI(I) 1	2	Tax(I) 1	2	Earn(I) 1	2
1	35	37	75	80	37	40	73	77
2	35	37	72	77	36	38	71	76
3	35	37	74	80	37	40	72	77
4	35	37	77	83	38	41	74	79
5	35	37	79	85	39	42	75	80
6	35	37	82	88	41	44	76	81
7	35	37	85	91	42	45	78	83
8	35	37	88	94	44	47	79	84
9	35	37	91	97	45	48	81	86
10	35	37	94	100	47	50	82	87
11	35	37	97	104	48	52	84	89
12	35	37	97	104	48	52	84	89
13	35	37	97	104	48	52	84	89
14	35	0	92	-20	46	-10	81	-10
15	35	0	92	-20	46	-10	81	-10
16	35	0	92	-20	46	-10	81	-10
17	35	0	92	-20	46	-10	81	-10
18	35	0	92	-20	46	-10	81	-10
19	35	0	92	-20	46	-10	81	-10
20	35	0	92	-20	46	-10	81	-10
21	35	0	92	-20	46	-10	81	-10
22	35	0	92	-20	46	-10	81	-10
23	35	0	92	-20	46	-10	81	-10
24	35	0	92	-20	46	-10	81	-10
25	35	0	92	-20	46	-10	81	-10

Table G-14. The Changes in Taxable Income, Taxes Paid, and
Earnings When the Depletion Allowance is .14

	DA(I)		TI(I)		Tax(I)		Earn(I)	
Facility	1	2	1	2	1	2	1	2
Time								
1	38	40	66	71	33	35	71	76
2	38	40	63	68	31	34	70	74
3	38	40	66	71	33	35	71	76
4	38	40	68	73	34	36	72	77
5	38	40	71	76	35	38	74	78
6	38	40	73	79	36	39	75	80
7	38	40	76	82	38	41	76	81
8	38	40	79	85	39	42	78	83
9	38	40	81	88	40	44	79	84
10	38	40	84	91	42	45	80	86
11	38	40	87	94	43	47	82	87
12	38	40	89	95	44	47	83	88
13	38	40	89	95	44	47	83	88
14	38	40	89	95	44	47	83	88
15	38	0	82	-20	41	-10	79	-10
16	38	0	82	-20	41	-10	79	-10
17	38	0	82	-20	41	-10	79	-10
18	38	0	82	-20	41	-10	79	-10
19	38	0	82	-20	41	-10	79	-10
20	38	0	82	-20	41	-10	79	-10
21	38	0	82	-20	41	-10	79	-10
22	38	0	82	-20	41	-10	79	-10
23	38	0	82	-20	41	-10	79	-10
24	38	0	82	-20	41	-10	79	-10
25	38	0	82	-20	41	-10	79	-10

457

450

Table G-15. The Changes in Taxable Income, Taxes Paid, and Earnings When the Depletion Allowance is .14

Facility Time	DA(I) 1	DA(I) 2	TI(I) 1	TI(I) 2	Tax(I) 1	Tax(I) 2	Earn(I) 1	Earn(I) 2
1	41	44	80	85	40	42	81	87
2	41	44	78	83	39	41	80	86
3	41	44	80	86	40	43	81	87
4	41	44	83	89	41	44	83	89
5	41	44	86	92	43	46	84	90
6	41	44	90	95	45	47	86	92
7	41	44	93	96	46	49	88	94
8	41	44	96	102	48	51	89	95
9	41	44	99	106	49	53	91	97
10	41	44	103	109	51	54	93	99
11	41	44	103	109	51	54	93	99
12	41	44	103	109	51	54	93	99
13	41	0	101	-20	50	-10	92	-10
14	41	0	101	-20	50	-10	92	-10
15	41	0	101	-20	50	-10	92	-10
16	41	0	101	-20	50	-10	92	-10
17	41	0	101	-20	50	-10	92	-10
18	41	0	101	-20	50	-10	92	-10
19	41	0	101	-20	50	-10	92	-10
20	41	0	101	-20	50	-10	92	-10
21	41	0	101	-20	50	-10	92	-10
22	41	0	101	-20	50	-10	92	-10
23	41	0	101	-20	50	-10	92	-10
24	41	0	101	-20	50	-10	92	-10
25	41	0	101	-20	50	-10	92	-10

Table G-16. The Changes in Taxable Income, Taxes Paid, and Earnings When the Depletion Allowance is .15

Facility	DA(I)		TI(I)		Tax (I)		Earn(I)	
Time	1	2	1	2	1	2	1	2
1	52	47	96	83	48	41	100	89
2	52	47	93	81	46	40	99	88
3	52	47	97	85	48	42	101	90
4	52	47	101	88	50	44	103	91
5	52	47	105	92	52	46	105	93
6	52	47	110	95	55	47	107	95
7	52	47	114	99	57	49	109	97
8	52	47	118	103	59	51	111	99
9	52	47	122	107	61	53	113	101
10	52	47	122	107	61	53	113	101
11	52	47	122	107	61	53	113	101
12	52	47	122	107	61	53	113	101
13	52	47	122	107	61	53	113	101
14	52	0	119	-20	60	-10	112	-10
15	52	0	119	-20	60	-10	112	-10
16	52	0	119	-20	60	-10	112	-10
17	52	0	119	-20	60	-10	112	-10
18	52	0	119	-20	60	-10	112	-10
19	52	0	119	-20	60	-10	112	-10
20	52	0	119	-20	60	-10	112	-10
21	52	0	119	-20	60	-10	112	-10
22	52	0	119	-20	60	-10	112	-10
23	52	0	119	-20	60	-10	112	-10
24	52	0	119	-20	60	-10	112	-10
25	52	0	119	-20	60	-10	112	-10

Table G-17. The Changes in Taxable Income, Taxes Paid, and Earnings When the Depletion Allowance is .16

Facility Time	DA(I) 1	2	TI(I) 1	2	Tax(I) 1	2	Earn(I) 1	2
1	47	50	74	79	37	39	84	90
2	47	50	72	77	36	38	83	89
3	47	50	75	80	37	40	85	90
4	47	50	78	83	39	41	86	92
5	47	50	81	87	40	43	88	94
6	47	50	84	90	42	45	89	95
7	47	50	87	94	43	47	91	97
8	47	50	91	91	45	48	93	99
9	47	50	94	97	47	50	94	101
10	47	50	97	101	48	51	96	102
11	47	50	97	103	48	51	96	102
12	47	50	97	103	48	51	96	-10
13	47	0	95	103	47	-10	95	-10
14	47	0	95	-20	47	-10	95	-10
15	47	0	95	-20	47	-10	95	-10
16	47	0	95	-20	47	-10	95	-10
17	47	0	95	-20	47	-10	95	-10
18	47	0	95	-20	47	-10	95	-10
19	47	0	95	-20	47	-10	95	-10
20	47	0	95	-20	47	-10	95	-10
21	47	0	95	-20	47	-10	95	-10
22	47	0	95	-20	47	-10	95	-10
23	47	0	95	-20	47	-10	95	-10
24	47	0	95	-20	47	-10	95	-10
25	47	0	95	-20	47	-10	95	-10

Table G-18. The Changes in Taxable Income, Taxes Paid, and Earnings When the Depletion Allowance is .17

Facility Time	DA(I) 1	2	TI(I) 1	2	Tax(I) 1	2	Earn(I) 1	2
1	50	53	71	76	35	38	86	91
2	50	53	69	74	34	37	85	90
3	50	53	72	77	36	38	86	92
4	50	53	75	80	37	40	88	93
5	50	53	78	84	39	42	89	95
6	50	53	82	87	41	43	91	97
7	50	53	85	91	42	45	93	99
8	50	53	88	95	44	47	94	101
9	50	53	92	99	46	49	96	103
10	50	53	94	100	47	50	97	103
11	50	53	94	100	47	50	97	103
12	50	0	92	-20	46	-10	96	-10
13	50	0	92	-20	46	-10	96	-10
14	50	0	92	-20	46	-10	96	-10
15	50	0	92	-20	46	-10	96	-10
16	50	0	92	-20	46	-10	96	-10
17	50	0	92	-20	46	-10	96	-10
18	50	0	92	-20	46	-10	96	-10
19	50	0	92	-20	46	-10	96	-10
20	50	0	92	-20	46	-10	96	-10
21	50	0	92	-20	46	-10	96	-10
22	50	0	92	-20	46	-10	96	-10
23	50	0	92	-20	46	-10	96	-10
24	50	0	92	-20	46	-10	96	-10
25	50	0	92	-20	46	-10	96	-10

Table G-19. The Changes in Taxable Income, Taxes Paid, and
Earnings When the Depletion Allowance is .18

Facility	DA(I) 1	DA(I) 2	TI(I) 1	TI(I) 2	Tax(I) 1	Tax(I) 2	Earn(I) 1	Earn(I) 2
Time								
1	53	56	62	67	31	33	84	90
2	53	56	60	65	30	32	83	89
3	53	56	63	68	31	36	85	90
4	53	56	66	71	33	35	86	92
5	53	56	69	75	34	37	88	94
6	53	56	73	78	36	39	90	95
7	53	56	76	82	38	41	91	97
8	53	56	79	85	39	42	93	99
9	53	56	83	89	41	44	95	101
10	53	56	85	91	42	45	96	102
11	53	56	85	91	42	45	96	102
12	53	56	85	91	42	45	96	102
13	53	0	82	-20	41	-10	94	-10
14	53	0	82	-20	41	-10	94	-10
15	53	0	82	-20	41	-10	94	-10
16	53	0	82	-20	41	-10	94	-10
17	53	0	82	-20	41	-10	94	-10
18	53	0	82	-20	41	-10	94	-10
19	53	0	82	-20	41	-10	94	-10
20	53	0	82	-20	41	-10	94	-10
21	53	0	82	-20	41	-10	94	-10
22	53	0	82	-20	41	-10	94	-10
23	53	0	82	-20	41	-10	94	-10
24	53	0	82	-20	41	-10	94	-10
25	53	0	82	-20	41	-10	94	-10

Table G-20. The Changes in Taxable Income, Taxes Paid, and Earnings When the Depletion Allowance is .19

Facility Time	DA(I) 1	DA(I) 2	TI(I) 1	TI(I) 2	Tax(I) 1	Tax(I) 2	Earn(I) 1	Earn(I) 2
1	56	59	54	58	27	29	83	88
2	56	59	51	56	25	28	82	87
3	56	59	54	59	27	29	83	89
4	56	59	57	62	28	31	85	90
5	56	59	60	65	30	32	86	92
6	56	59	63	69	31	34	88	94
7	56	59	67	72	33	36	90	95
8	56	59	70	76	35	38	91	97
9	56	59	73	79	36	39	93	99
10	56	59	76	82	38	41	94	100
11	56	59	76	82	38	41	94	100
12	56	59	76	82	38	41	94	100
13	56	0	71	-20	35	-10	92	-10
14	56	0	71	-20	35	-10	92	-10
15	56	0	71	-20	35	-10	92	-10
16	56	0	71	-20	35	-10	92	-10
17	56	0	71	-20	35	-10	92	-10
18	56	0	71	-20	35	-10	92	-10
19	56	0	71	-20	35	-10	92	-10
20	56	0	71	-20	35	-10	92	-10
21	56	0	71	-20	35	-10	92	-10
22	56	0	71	-20	35	-10	92	-10
23	56	0	71	-20	35	-10	92	-10
24	56	0	71	-20	35	-10	92	-10
25	56	0	71	-20	35	-10	92	-10

Table G-21. The Changes in Taxable Income, Taxes Paid, and Earnings When the Depletion Allowance is .20

Facility Time	DA(I) 1	2	TI(I) 1	2	Tax(I) 1	2	Earn(I) 1	2
1	59	63	45	48	22	24	82	87
2	59	63	42	46	21	23	80	86
3	59	63	45	49	22	24	82	88
4	59	63	58	52	24	26	83	89
5	59	63	51	55	25	27	85	91
6	59	63	55	58	27	29	87	92
7	59	63	58	62	29	31	88	94
8	59	63	61	65	30	32	90	96
9	59	63	64	69	32	34	91	98
10	59	63	68	72	34	36	93	99
11	59	63	68	72	34	36	93	99
12	59	63	68	72	34	36	93	99
13	59	0	61	-20	30	-10	90	-10
14	59	0	61	-20	30	-10	90	-10
15	59	0	61	-20	30	-10	90	-10
16	59	0	61	-20	30	-10	90	-10
17	59	0	61	-20	30	-10	90	-10
18	59	0	61	-20	30	-10	90	-10
19	59	0	61	-20	30	-10	90	-10
20	59	0	61	-20	30	-10	90	-10
21	59	0	61	-20	30	-10	90	-10
22	59	0	61	-20	30	-10	90	-10
23	59	0	61	-20	30	-10	90	-10
24	59	0	61	-20	30	-10	90	-10
25	59	0	61	-20	30	-10	90	-10

Table G-22. The Changes in Taxable Income, Taxes Paid, and Earnings When the Depletion Allowance is .21

Facility Time	DA(I) 1	DA(I) 2	TI(I) 1	TI(I) 2	Tax(I) 1	Tax(I) 2	Earn(I) 1	Earn(I) 2
1	62	66	59	63	29	31	92	98
2	62	66	57	61	28	30	91	97
3	62	66	60	65	30	32	92	99
4	62	66	64	68	32	34	94	100
5	62	66	67	72	33	36	96	102
6	62	66	71	76	35	38	98	104
7	62	66	75	80	37	40	100	106
8	62	66	78	84	39	42	101	108
9	62	66	82	87	41	43	103	110
10	62	66	82	87	41	43	103	110
11	62	66	82	87	41	43	103	110
12	62	0	80	-20	40	-10	102	-10
13	62	0	80	-20	40	-10	102	-10
14	62	0	80	-20	40	-10	102	-10
15	62	0	80	-20	40	-10	102	-10
16	62	0	80	-20	40	-10	102	-10
17	62	0	80	-20	40	-10	102	-10
18	62	0	80	-20	40	-10	102	-10
19	62	0	80	-20	40	-10	102	-10
20	62	0	80	-20	40	-10	102	-10
21	62	0	80	-20	40	-10	102	-10
22	62	0	80	-20	40	-10	102	-10
23	62	0	80	-20	40	-10	102	-10
24	62	0	80	-20	40	-10	102	-10
25	62	0	80	-20	40	-10	102	-10

Table G-23. The Changes in Taxable Income, Taxes Paid, and
Earnings When the Depletion Allowance is .22

Facility Time	DA(I) 1	2	TI(I) 1	2	Tax(I) 1	2	Earn(I) 1	2
1	77	69	71	61	35	30	113	100
2	77	69	69	60	34	30	112	99
3	77	69	74	64	37	32	114	101
4	77	69	78	68	39	34	116	103
5	77	69	83	72	41	36	119	105
6	77	69	87	76	43	38	121	107
7	77	69	92	80	46	40	123	109
8	77	69	97	85	48	42	126	112
9	77	69	97	85	48	42	126	112
10	77	69	97	85	48	42	126	112
11	77	69	97	85	48	42	126	112
12	77	0	96	-20	48	-10	125	-10
13	77	0	96	-20	48	-10	125	-10
14	77	0	96	-20	48	-10	125	-10
15	77	0	96	-20	48	-10	125	-10
16	77	0	96	-20	48	-10	125	-10
17	77	0	96	-20	48	-10	125	-10
18	77	0	96	-20	48	-10	125	-10
19	77	0	96	-20	48	-10	125	-10
20	77	0	96	-20	48	-10	125	-10
21	77	0	96	-20	48	-10	125	-10
22	77	0	96	-20	48	-10	125	-10
23	77	0	96	-20	48	-10	125	-10
24	77	0	96	-20	48	-10	125	-10
25	77	0	96	-20	48	-10	125	-10

Table G-24. The Changes in Taxable Income, Taxes Paid, and
Earnings When the Depletion Allowance is .23

Facility Time	DA(I) 1	2	TI(I) 1	2	Tax(I) 1	2	Earn(I) 1	2
1	68	72	53	57	26	28	95	101
2	68	72	51	55	25	27	94	100
3	68	72	54	59	27	29	95	102
4	68	72	58	63	29	31	97	104
5	68	72	62	67	31	33	99	106
6	68	72	66	71	33	35	101	108
7	68	72	69	74	34	37	103	109
8	68	72	73	79	36	39	105	112
9	68	72	76	81	38	40	106	113
10	68	72	76	81	38	40	106	113
11	68	72	76	81	38	40	105	-10
12	68	72	74	-20	37	-10	105	-10
13	68	0	74	-20	37	-10	105	-10
14	68	0	74	-20	37	-10	105	-10
15	68	0	74	-20	37	-10	105	-10
16	68	0	74	-20	37	-10	105	-10
17	68	0	74	-20	37	-10	105	-10
18	68	0	74	-20	37	-10	105	-10
19	68	0	74	-20	37	-10	105	-10
20	68	0	74	-20	37	-10	105	-10
21	68	0	74	-20	37	-10	105	-10
22	68	0	74	-20	37	-10	105	-10
23	68	0	74	-20	37	-10	105	-10
24	68	0	74	-20	37	-10	105	-10
25	68	0	74	-20	37	-10	105	-10

Table G-25. The Changes in Taxable Income, Taxes Paid, and
Earnings When the Depletion Allowance is .24

Facility Time	DA(I) 1	2	TI(I) 1	2	Tax(I) 1	2	Earn(I) 1	2
1	71	75	50	54	25	27	96	102
2	71	75	48	52	24	26	95	101
3	71	75	52	56	26	28	97	103
4	71	75	55	60	27	30	99	105
5	71	75	59	64	29	32	101	107
6	71	75	63	68	31	34	103	109
7	71	75	67	72	33	36	105	111
8	71	75	71	76	35	38	107	113
9	71	75	73	78	36	39	108	114
10	71	75	73	78	36	39	107	114
11	71	0	71	-20	35	-10	107	-10
12	71	0	71	-20	35	-10	107	-10
13	71	0	71	-20	35	-10	107	-10
14	71	0	71	-20	35	-10	107	-10
15	71	0	71	-20	35	-10	107	-10
16	71	0	71	-20	35	-10	107	-10
17	71	0	71	-20	35	-10	107	-10
18	71	0	71	-20	35	-10	107	-10
19	71	0	71	-20	35	-10	107	-10
20	71	0	71	-20	35	-10	107	-10
21	71	0	71	-20	35	-10	107	-10
22	71	0	71	-20	35	-10	107	-10
23	71	0	71	-20	35	-10	107	-10
24	71	0	71	-20	35	-10	107	-10
25	71	0	71	-20	35	-10	107	-10

Table G-26. The Changes in Taxable Income, Taxes Paid, and
Earnings When the Depletion Allowance is .25

Facility	DA(I)		TI(I)		Tax(I)		Earn(I)	
Time	1	2	1	2	1	2	1	2
1	74	78	41	45	20	22	95	101
2	74	78	39	43	19	21	94	100
3	74	78	43	47	21	23	96	102
4	74	78	46	51	23	25	97	104
5	74	78	50	54	25	27	99	105
6	74	78	54	59	27	29	101	108
7	74	78	58	62	29	31	103	109
8	74	78	62	67	31	33	105	112
9	74	78	64	69	32	34	106	113
10	74	78	64	69	32	34	106	113
11	74	78	64	69	32	34	106	113
12	74	0	61	-20	30	-10	105	-10
13	74	0	61	-20	30	-10	105	-10
14	74	0	61	-20	30	-10	105	-10
15	74	0	61	-20	30	-10	105	-10
16	74	0	61	-20	30	-10	105	-10
17	74	0	61	-20	30	-10	105	-10
18	74	0	61	-20	30	-10	105	-10
19	74	0	61	-20	30	-10	105	-10
20	74	0	61	-20	30	-10	105	-10
21	74	0	61	-20	30	-10	105	-10
22	74	0	61	-20	30	-10	105	-10
23	74	0	61	-20	30	-10	105	-10
24	74	0	61	-20	30	-10	105	-10
25	74	0	61	-20	30	-10	105	-10

Table G-27. The Changes in Taxable Income, Taxes Paid, and
Earnings When the Depletion Allowance is .26

Facility Time	DA(I) 1	2	TI(I) 1	2	Tax(I) 1	2	Earn(I) 1	2
1	77	81	33	36	16	18	94	99
2	77	81	31	34	15	17	93	98
3	77	81	34	38	17	19	94	100
4	77	81	37	41	18	20	96	102
5	77	81	41	45	20	22	98	104
6	77	81	45	49	22	24	100	106
7	77	81	49	53	24	26	102	108
8	77	81	52	57	26	28	103	110
9	77	81	55	60	27	30	105	111
10	77	81	55	60	27	30	105	111
11	77	81	55	60	27	30	105	111
12	77	0	50	-20	25	-10	102	-10
13	77	0	50	-20	25	-10	102	-10
14	77	0	50	-20	25	-10	102	-10
15	77	0	50	-20	25	-10	102	-10
16	77	0	50	-20	25	-10	102	-10
17	77	0	50	-20	25	-10	102	-10
18	77	0	50	-20	25	-10	102	-10
19	77	0	50	-20	25	-10	102	-10
20	77	0	50	-20	25	-10	102	-10
21	77	0	50	-20	25	-10	102	-10
22	77	0	50	-20	25	-10	102	-10
23	77	0	50	-20	25	-10	102	-10
24	77	0	50	-20	25	-10	102	-10
25	77	0	50	-20	25	-10	102	-10

Table G-28. The Changes in Taxable Income, Taxes Paid, and Earnings When the Depletion Allowance is .27

Facility Time	DA(I) 1	2	TI(I) 1	2	Tax(I) 1	2	Earn(I) 1	2
1	80	85	24	26	12	13	92	98
2	80	85	22	24	11	12	91	97
3	80	85	25	27	12	13	93	99
4	80	85	29	31	14	15	95	101
5	80	85	32	35	16	17	96	103
6	80	85	36	39	18	19	98	105
7	80	85	39	43	19	21	100	107
8	80	85	43	47	21	23	102	109
9	80	85	47	50	23	25	104	110
10	80	85	47	50	23	25	104	110
11	80	85	47	50	23	25	104	110
12	80	0	40	-20	20	-10	100	-10
13	80	0	40	-20	20	-10	100	-10
14	80	0	40	-20	20	-10	100	-10
15	80	0	40	-20	20	-10	100	-10
16	80	0	40	-20	20	-10	100	-10
17	80	0	40	-20	20	-10	100	-10
18	80	0	40	-20	20	-10	100	-10
19	80	0	40	-20	20	-10	100	-10
20	80	0	40	-20	20	-10	100	-10
21	80	0	40	-20	20	-10	100	-10
22	80	0	40	-20	20	-10	100	-10
23	80	0	40	-20	20	-10	100	-10
24	80	0	40	-20	20	-10	100	-10
25	80	0	40	-20	20	-10	100	-10

Table G-29. The Behavior of the Present Value of After-Tax Earnings With Changes In The Depletion Allowance.

Time/DA	.00	.01	.02	.03
1	84.55	98.18	89.09	90.91
2	158.93	184.96	167.60	171.90
3	290.82	341.23	240.48	246.28
4	349.80	410.78	308.10	315.27
5	405.12	475.69	370.81	379.22
6	455.92	535.73	428.95	439.62
7	503.04	591.71	482.83	495.04
8	547.15	643.87	532.75	546.36
9	588.02	692.45	579.40	593.85
10	625.87	736.99	622.58	638.58
11	660.92	778.71	662.54	679.59
12	693.36	816.65	699.82	717.82
13	723.38	851.15	734.29	752.87
14	750.67	882.51	765.63	784.73
15	775.48	911.02	794.11	813.70
16	782.41	936.94	820.01	821.97
17	788.70	960.50	827.33	829.49
18	794.42	981.92	833.99	836.32
19	799.63	1001.39	840.04	842.54
20	804.35	1019.09	845.54	842.18
21	808.65	1035.18	850.54	853.32
22	812.56	1049.81	855.08	857.99
23	816.12	1063.11	859.21	862.23
24	819.35	1075.20	862.97	866.09
25	823.67	1084.83	866.38	869.00

Time/DA	.04	.05	.06	.07
1	89.09	86.36	84.55	101.82
2	167.60	162.40	158.93	191.07
3	240.48	233.02	227.30	274.47
4	308.10	299.27	290.82	352.33
5	370.81	360.74	349.80	424.36
6	428.95	417.19	405.12	490.97
7	482.83	469.53	455.92	553.06
8	533.22	508.52	503.04	610.91
9	579.87	563.90	547.15	664.34
10	623.05	605.53	588.02	714.08
11	663.01	644.44	625.87	759.99
12	700.29	680.76	660.92	802.37
13	734.26	714.36	693.36	840.90
14	766.09	744.91	723.38	815.92
15	794.58	772.68	750.67	886.45
16	820.48	797.92	775.48	896.03
17	827.60	804.85	782.01	904.73
18	834.07	811.14	787.94	912.65
19	839.96	816.87	793.34	919.84
20	845.31	822.07	798.25	926.38
21	850.18	826.80	802.70	932.33
22	854.60	831.10	806.76	937.73
23	858.62	835.01	810.44	942.65
24	862.27	838.56	813.79	947.11
25	865.60	841.79	816.84	951.18

Time/DA	.08	.09	.10	.11
1	117.27	106.36	108.18	106.36
2	222.23	200.58	204.88	200.58
3	319.15	288.48	294.28	288.48
4	409.99	369.76	378.29	369.76
5	493.82	445.51	455.91	445.51
6	572.28	516.07	527.60	516.07
7	644.63	581.24	594.31	581.24
8	711.81	641.89	656.35	641.89
9	744.58	698.29	713.61	698.29
10	832.79	750.34	766.81	750.34
11	886.07	798.71	816.23	798.71
12	934.50	842.68	861.16	842.68
13	978.53	882.66	902.00	882.66
14	1018.56	894.77	914.64	894.77
15	1054.94	905.78	926.13	905.78
16	1088.02	915.79	936.58	915.79
17	1118.10	924.87	946.07	924.89
18	1145.43	933.17	954.71	933.17
19	1170.29	940.69	962.56	940.69
20	1192.88	947.63	969.69	947.53
21	1213.42	953.74	976.18	953.74
22	1232.09	959.39	982.07	959.39
23	1249.07	964.53	987.43	964.53
24	1264.50	969.20	992.31	969.20
25	1278.53	973.45	996.74	973.45

Time/DA	.12	.13	.14	.15
1	104.55	101.82	119.09	136.36
2	197.11	191.90	225.70	258.21
3	282.76	276.05	324.12	372.88
4	362.67	353.91	416.33	478.06
5	436.56	426.56	500.78	576.17
6	504.86	493.73	579.80	667.61
7	569.01	555.82	653.70	752.28
8	627.79	614.14	721.34	831.12
9	682.92	667.57	784.53	904.07
10	733.81	717.31	843.52	970.38
11	781.13	763.57	897.15	1036.67
12	824.14	806.27	945.90	1085.47
13	863.25	845.09	961.54	1135.29
14	874.83	880.37	975.76	1180.59
15	885.37	890.43	988.69	1212.76
16	894.94	899.57	1000.44	1259.19
17	903.65	907.88	1011.12	1293.22
18	911.56	915.43	1020.83	1324.16
19	918.76	922.30	1029.66	1352.28
20	925.30	928.54	1037.69	1377.85
21	931.24	934.22	1044.99	1401.09
22	936.65	939.38	1051.62	1422.22
23	941.56	944.07	1057.65	1441.43
24	946.03	948.33	1063.13	1458.89
25	950.09	952.21	1068.12	1474.77

Time/DA	.16	.17	.18	.19
1	123.64	126.36	123.64	121.82
2	235.21	239.59	235.21	230.91
3	338.14	344.77	338.14	332.34
4	433.76	442.44	433.76	425.91
5	523.17	532.48	523.17	512.84
6	605.02	616.58	605.58	594.12
7	681.48	695.10	682.05	669.04
8	752.86	767.40	753.42	738.55
9	818.59	834.84	819.58	803.44
10	879.51	896.52	880.50	862.81
11	934.89	952.60	935.87	916.79
12	985.23	970.76	986.22	965.86
13	1001.45	987.27	1002.15	981.50
14	1016.20	1002.28	1016.66	995.72
15	1029.60	1015.93	1029.80	1008.65
16	1041.79	1028.33	1041.77	1020.40
17	1052.82	1039.61	1052.65	1031.08
18	1062.94	1049.86	1062.54	1040.80
19	1072.10	1059.18	1071.59	1049.63
20	1080.42	1067.66	1079.71	1057.65
21	1087.99	1075.36	1087.14	1064.95
22	1094.87	1082.36	1093.90	1071.58
23	1101.12	1088.73	1106.04	1077.61
24	1106.81	1094.51	1105.63	1083.10
25	1111.98	1099.77	1110.70	1088.06

Time/DA	.20	.21	.22	.23
1	120.00	137.27	155.45	141.82
2	226.61	260.41	295.95	269.09
3	326.54	374.61	425.93	387.05
4	418.74	479.80	546.82	497.01
5	504.43	577.90	659.21	598.84
6	584.02	669.35	763.64	693.68
7	657.92	754.02	860.62	781.43
8	726.49	832.39	951.13	863.06
9	790.11	904.91	1033.40	938.13
10	849.10	970.84	1180.20	1006.37
11	902.72	1030.77	1176.19	1068.41
12	951.47	1050.85	1238.01	1089.12
13	966.54	1069.10	1294.20	1107.95
14	980.23	1085.69	1345.29	1125.06
15	992.68	1100.77	1391.73	1140.62
16	1003.99	1114.46	1433.95	1154.77
17	1014.29	1126.94	1472.33	1167.63
18	1023.63	1138.27	1507.22	1179.32
19	1032.14	1148.58	1538.94	1189.95
20	1039.86	1157.94	1567.78	1199.61
21	1046.89	1166.45	1594.00	1208.39
22	1053.28	1174.19	1617.83	1216.38
23	1059.09	1181.23	1639.49	1223.66
24	1064.37	1187.62	1659.19	1230.24
25	1069.17	1193.44	1677.10	1236.24

Time/DA	.24	.25	.26	.27
1	143.64	141.82	139.09	137.27
2	272.56	269.09	264.71	260.41
3	392.77	387.80	380.41	375.36
4	504.10	497.76	488.33	481.91
5	607.08	598.97	588.92	580.64
6	704.32	693.81	682.06	672.09
7	793.61	781.56	768.78	757.27
8	876.65	863.19	848.55	836.58
9	952.99	938.26	922.35	909.58
10	1022.39	1006.25	989.43	975.83
11	1045.87	1068.54	1050.42	1036.12
12	1067.22	1089.25	1070.49	1055.56
13	1086.62	1108.08	1088.74	1073.23
14	1104.27	1125.19	1105.33	1089.29
15	1120.31	1140.75	1120.41	1103.89
16	1134.89	1154.90	1134.12	1117.17
17	1148.14	1167.76	1146.59	1129.24
18	1160.19	1179.45	1157.92	1140.21
19	1171.15	1190.08	1168.22	1150.18
20	1181.11	1199.74	1177.58	1159.25
21	1190.16	1208.52	1186.10	1167.49
22	1198.39	1216.51	1193.83	1174.98
23	1205.88	1223.77	1200.87	1181.80
24	1212.68	1230.37	1207.27	1187.89
25	1218.86	1236.37	1213.08	1193.62

APPENDIX H

Changes In The Dividend Rate

Another parameter that was allowed to vary was the
dividends paid by the corporation to stockholders. This
modification was undertaken since it would illustrate the
effects of varying dividend policy on the objective function,
and secondly, since it adds realism to the model in that
many corporations are subject to the stockholders influence
and thus have to consider dividend policy when considering
overall corporate strategy.

The general effect is that as the percentage paid
in dividends increases, first the debt level is decreased,
and eventually, if the dividend rate is increased sufficiently, to the point where investments are deferred to
the overall detriment of the firm. In Table H-1 we have
indicated changes in the dividend payout rate from .01 to
.30. It is shown that as we increase the payout rate, the
present value of the after-tax earnings decreases, and also
investment decreases over time.

This analysis can then be used by management to determine an optimal dividend payout ratio so that over the planning horizon, the stockholders are kept happy with their returns yet the dividend rate is such that growth of earnings and investment over time are unhampered.

Table H-1.   Changes in the Dividend Rates and Cumulative
             Dividends

DR Time	.01	.02	.03	.04	.05
1	1.70	3.40	5.10	6.80	8.50
2	3.24	6.47	9.71	12.92	16.14
3	4.66	9.33	13.99	18.63	23.28
4	6.00	11.99	17.99	23.93	29.91
5	7.23	14.46	21.70	28.87	36.09
6	8.38	16.76	25.13	33.45	41.81
7	9.45	18.89	28.34	37.70	47.13
8	10.43	20.86	31.29	41.64	52.05
9	11.33	22.65	33.97	45.22	56.52
10	11.64	23.27	34.90	46.45	60.59
11	11.92	23.83	35.74	47.57	61.99
12	12.17	24.34	36.51	48.59	63.27
13	12.41	24.80	37.20	49.52	64.42
14	12.62	25.22	37.83	50.36	65.48
15	12.81	25.60	38.41	51.13	66.44
16	12.98	25.95	38.93	51.82	67.31
17	13.14	26.27	39.40	52.46	68.10
18	13.28	26.56	39.84	53.03	68.82
19	13.41	26.82	40.23	53.56	69.47
20	13.53	27.06	40.58	54.03	70.07
21	13.64	27.27	40.91	54.46	70.61
22	13.74	27.47	41.20	54.86	71.10
23	13.83	27.65	41.47	55.22	71.54
24	13.91	27.81	41.72	55.54	71.95
25	13.98	27.96	41.94	55.84	72.32

474

DR Time	.06	.07	.08	.09	.10	.11
1	10.20	13.24	15.13	15.30	17.00	18.70
2	19.37	25.27	28.88	29.06	32.29	35.52
3	27.89	36.42	41.62	41.84	46.49	51.14
4	35.80	46.79	53.48	53.70	59.67	65.64
5	43.18	56.44	64.51	64.71	71.90	79.09
6	50.05	65.41	74.76	74.92	83.25	91.57
7	56.43	73.78	84.24	84.44	93.77	103.15
8	62.33	81.39	92.94	93.30	103.57	113.92
9	67.70	88.31	100.84	101.35	112.51	123.77
10	72.58	94.60	108.03	108.67	120.65	132.71
11	74.27	100.31	114.56	111.20	123.45	135.80
12	75.80	105.51	120.50	113.49	126.00	138.60
13	77.19	110.23	125.90	115.58	128.32	141.15
14	78.45	114.53	130.81	117.47	130.43	143.47
15	79.60	118.43	135.27	119.20	132.34	145.58
16	80.64	121.98	139.33	120.76	134.08	147.49
17	81.59	125.21	143.01	122.19	135.67	149.23
18	82.46	128.14	146.37	123.48	137.10	150.81
19	83.24	130.81	149.42	124.66	138.41	152.25
20	83.96	133.23	152.19	125.73	139.60	153.56
21	84.60	135.44	154.70	126.70	140.68	154.75
22	85.19	137.44	156.99	127.59	141.66	155.83
23	85.73	139.26	159.08	128.39	142.56	156.81
24	86.22	140.92	160.97	129.12	143.37	157.71
25	86.66	142.42	162.69	129.79	144.11	158.52

DR Time	.12	.13	.14	.15	.16
1	20.40	22.10	23.80	28.36	27.20
2	38.75	41.98	54.02	54.02	51.66
3	55.79	60.44	77.92	77.92	74.38
4	71.61	77.57	83.54	100.05	95.47
5	86.28	93.47	108.22	120.54	115.05
6	99.90	108.22	121.90	139.59	133.20
7	112.52	121.90	134.57	157.21	149.95
8	124.22	134.57	146.21	173.52	165.47
9	134.96	146.21	156.78	188.34	179.79
10	144.72	156.78	160.43	201.82	192.81
11	148.09	160.43	163.74	214.07	197.29
12	151.15	163.74	166.75	225.20	201.37
13	153.93	166.75	169.49	235.33	205.08
14	156.46	169.49	171.98	244.53	208.45
15	158.75	171.98	174.25	252.90	211.51
16	160.84	174.25	176.30	260.50	214.30
17	162.74	176.30	178.17	267.42	216.83
18	164.47	178.17	179.81	273.70	219.13
19	166.04	179.87	181.42	279.42	221.23
20	167.47	181.42	182.83	284.61	223.13
21	168.76	182.83	184.10	289.34	224.86
22	169.94	184.10	185.27	293.63	226.43
23	171.01	185.27	186.32	297.53	227.86
24	171.99	186.32	187.28	301.08	229.16
25	172.87	187.28	187.65	304.31	230.34

DR Time	.17	.18	.19	.20	.25	.30
1	28.90	30.60	32.30	34.00	42.50	51.00
2	54.89	58.10	61.35	64.58	80.72	96.62
3	79.03	83.68	88.33	92.98	116.03	138.77
4	101.44	107.29	113.25	119.21	148.65	177.91
5	122.24	129.08	136.25	143.42	178.92	213.86
6	141.33	149.30	157.59	165.89	206.86	247.22
7	159.04	168.05	177.39	186.72	232.65	278.16
8	175.46	185.43	195.73	206.04	236.55	306.57
9	190.67	201.54	212.74	223.93	278.71	332.91
10	204.50	216.18	228.19	240.20	299.05	357.20
11	217.07	229.49	242.24	254.99	317.54	379.38
12	221.40	234.08	247.09	260.09	334.35	399.55
13	225.34	238.25	251.49	264.73	340.14	406.50
14	228.93	242.05	255.49	268.94	345.41	412.82
15	232.18	245.49	259.13	272.77	350.20	418.57
16	235.14	248.63	262.44	276.25	354.55	423.79
17	237.83	251.48	265.45	279.42	358.50	428.54
18	240.28	254.07	268.18	282.29	365.37	432.86
19	242.50	256.42	270.67	284.91	368.35	436.78
20	244.52	258.56	272.92	287.29	371.05	440.35
21	246.36	260.51	274.98	289.45	373.50	443.59
22	248.03	262.28	276.85	291.42	375.74	446.54
23	249.55	263.88	278.54	293.20	377.77	449.22
24	250.93	265.35	280.07	294.83	379.61	451.66
25	252.19	266.67	281.49	296.30	381.46	453.87

APPENDIX I

Changes In The Sales and
Administrative Costs

A further parameter that was allowed to vary was
sales and administrative expenses. These changes would be
exhibited by changes in the cost parameters. However, sales
and administrative expenses were such an insignificant proportion of the total costs involved in operating a reservoir
or well that changes in these parameters demonstrated no
effect on the objective function or on the optimal investment policy.

APPENDIX J

Changes In The Tax Rate

Finally, as a last test, we allow for a change in the environment in which the firm operates, namely we allowed for changes in the tax rate to vary between the high of .50 to a low of .35. While the firm itself has no control over the rate of taxes it pays, the amount it does pay will effect the optimal investment policy and the after-tax earnings function of the firm.

As the depletion allowance greatly increased the after-tax earnings function, the higher its level, so changes in the tax rate greatly effected the after-tax earnings. Only now the lower tax was as a subsidy and thus the lower the tax rate, the less taxes paid, hence the greater after-tax earnings and hence more investment. This is clearly seen in Table J-1.which shows the changes in taxable income with changes in the tax rate.

The introduction of the parameter allows us to consider other environmental changes which might affect the firm, the after-tax earnings function and hence the optimal investment policy. Similar analysis would hold for such items as the tax credit on new investment, special tariffs granted select industries, price supports, import controls and other similar factors. The model can be used to show the sensitivity of changes in these parameters to potential investment and thus be used as a decision tool for effective management.

Table J-1. Changes in Taxable Income with Changes in the
Tax Rate

Facility	Taxable Income		Tax(I) .49		.48	
Time	1	2	1	2	1	2
1	40	43	19	21	19	21
2	38	41	18	21	18	20
3	42	46	20	22	20	21
4	46	50	22	24	21	24
5	50	54	24	26	23	26
6	54	58	26	28	25	28
7	58	62	28	30	27	30
8	62	67	30	32	29	32
9	63	67	30	32	29	32
10	63	67	30	32	29	32
11	61	-20	29	-9	29	-9
12	61	-20	29	-9	29	-9
13	61	-20	29	-9	29	-9
14	61	-20	29	-9	29	-9
15	61	-20	29	-9	29	-9
16	61	-20	29	-9	29	-9
17	61	-20	29	-9	29	-9
18	61	-20	29	-9	29	-9
19	61	-20	29	-9	29	-9
20	61	-20	29	-9	29	-9
21	61	-20	29	-9	29	-9
22	61	-20	29	-9	29	-9
23	61	-20	29	-9	29	-9
24	61	-20	29	-9	29	-9
25	61	-20	29	-9	29	-9

Rate	.47		.46		.45		.44		.43	
Facility	1	2	1	2	1	2	1	2	1	2
Time										
1	18	20	18	19	17	19	17	18	17	18
2	17	19	17	18	17	18	16	18	16	18
3	19	21	19	21	18	20	18	20	18	19
4	21	23	21	23	20	22	20	21	19	21
5	23	25	23	24	22	24	21	23	21	23
6	25	27	24	26	24	26	23	25	23	25
7	27	29	26	28	26	28	25	27	24	27
8	29	31	28	30	28	30	27	29	27	28
9	29	31	28	30	28	30	27	29	27	28
10	29	31	28	30	28	30	27	29	27	28
11	28	-9	28	-9	27	-8	26	-8	26	-8
12	28	-9	28	-9	27	-8	26	-8	26	-8
13	28	-9	28	-9	27	-8	26	-8	26	-8
14	28	-9	28	-9	27	-8	26	-8	26	-8
15	28	-9	28	-9	27	-8	26	-8	26	-8
16	28	-9	28	-9	27	-8	26	-8	26	-8
17	28	-9	28	-9	27	-8	26	-8	26	-8
18	28	-9	28	-9	27	-8	26	-8	26	-8
19	28	-9	28	-9	27	-8	26	-8	26	-8
20	28	-9	28	-9	27	-8	26	-8	26	-8
21	28	-9	28	-9	27	-8	26	-8	26	-8
22	28	-9	28	-9	27	-8	26	-8	26	-8
23	28	-9	28	-9	27	-8	26	-8	26	-8
24	28	-9	28	-9	27	-8	26	-8	26	-8
25	28	-9	28	-9	27	-8	26	-8	26	-8

Rate	.42		.41		.40		.39		.38	
Facility	1	2	1	2	1	2	1	2	1	2
Time										
1	16	17	16	17	16	17	15	16	15	16
2	16	17	15	16	15	16	14	16	14	15
3	17	18	16	18	16	18	16	17	15	17
4	18	20	18	20	18	20	17	19	17	19
5	20	22	20	21	20	21	19	21	19	20
6	22	25	21	23	21	23	21	23	20	22
7	24	26	23	25	23	25	23	24	22	23
8	26	27	25	26	25	26	24	26	23	25
9	26	27	25	26	25	26	24	26	23	25
10	25	27	25	26	25	26	24	26	23	25
11	25	-8	25	-8	24	-8	23	-7	23	-7
12	25	-8	25	-8	24	-8	23	-7	23	-7
13	25	-8	25	-8	24	-8	23	-7	23	-7
14	25	-8	25	-8	24	-8	23	-7	23	-7
15	25	-8	25	-8	24	-8	23	-7	23	-7
16	25	-8	25	-8	24	-8	23	-7	23	-7
17	25	-8	25	-8	24	-8	23	-7	23	-7
18	25	-8	25	-8	24	-8	23	-7	23	-7
19	25	-8	25	-8	24	-8	23	-7	23	-7
20	25	-8	25	-8	24	-8	23	-7	23	-7
21	25	-8	25	-8	24	-8	23	-7	23	-7
22	25	-8	25	-8	24	-8	23	-7	23	-7
23	25	-8	25	-8	24	-8	23	-7	23	-7
24	25	-8	25	-8	24	-8	23	-7	23	-7
25	25	-8	25	-8	24	-8	23			

Rate	.37		.36		.35	
Facility	1	2	1	2	1	2
Time						
1	14	15	14	15	13	15
2	14	15	13	15	13	14
3	15	17	15	16	14	16
4	17	18	16	17	16	17
5	18	19	17	19	17	18
6	20	21	19	21	19	20
7	21	23	21	22	20	22
8	23	24	22	24	22	23
9	23	24	22	24	22	23
10	23	24	22	24	22	23
11	22	-7	21	-7	21	-6
12	22	-7	21	-7	21	-6
13	22	-7	21	-7	21	-6
14	22	-7	21	-7	21	-6
15	22	-7	21	-7	21	-6
16	22	-7	21	-7	21	-6
17	22	-7	21	-7	21	-6
18	22	-7	21	-7	21	-6
19	22	-7	21	-7	21	-6
20	22	-7	21	-7	21	-6
21	22	-7	21	-7	21	-6
22	22	-7	21	-7	21	-6
23	22	-7	21	-7	21	-6
24	22	-7	21	-7	21	-6
25	22	-7	21	-7	21	-6

BIBLIOGRAPHY

BIBLIOGRAPHY

Books

1. Allen, R. D. G., Mathematical Analysis for Economist, (London, The MacMillan Co., 1938).

2. Aris, R.,Discrete Dynamic Programming, (New York, Blais- dell Press, 1964).

3. Arrow, K. J., S. Karlin, and H. Scarf, Studies In the Mathematical Theory of Inventory and Production, Stanford, California, Stanford University Press, 1958).

4. Beckman, M. J., Dynamic Programming of Economic Decisions, (New York, Akademie-Verlay, 1968).

5. Bellman, R., Dynamic Programming, (Princeton, N.J., Princeton University Press, 1957).

6. Bellman, R. and S. Dreyfus, Applied Dynamic Programming, (Princeton, N.J., Princeton University Press, 1962).

7. Bellman, R., Adaptive Control Processes, A Guided Tour, (Princeton, N.J., Princeton University Press, 1961).

8. Buckley, S. E. (ed.), Petroleum Conservation, (New York, American Petroleum Institute, 1951).

9. Calhoun, J. C., Fundamentals of Reservoir Engineering, (Norman, Oklahoma, The University of Oklahoma Press, 1953).

10. Charnes, A. and W. W. Cooper, Management Models and Industrial Applications of Linear Programming Volumes I and II, (New York, John Wiley & Sons Inc., 1961).

11. Charnes, A., W. W. Cooper, and A. Henderson, An Introduc- tion to Linear Programming, (New York, John Wiley & Sons Inc., 1953).

12. Chorafas, D. N., Control Systems Functions and Programming Approaches,Volumes A and B, (New York, Academic Press,(1966.)

13. Dean, J., Capital Budgeting, (New York, John Wiley & Sons Inc., 1951).

14. Dean, J., Managerial Economics, (Englewood Cliffs, N.J., Prentice-Hall Inc.,(1951.)

15. Di Rouaferrera, G. M., Operations Research Models for Business and Industry, (Cincinnati, Ohio, South-western Press, 1964.)

16. Dorfman, R. P., P. A. Samuelson, and R. Solow, Linear Programming and Economic Analysis, (New York, McGraw-Hill Inc., 1958).

17. Dreyfus, S., Dynamic Programming and The Calculus of Variations, (New York, Academic Press, 1965).

18. Fan, L. T. and W. S. Wang, The Discrete Maximum Principle: A Study of Multistage Systems Optimization, (New York, John Wiley and Sons Inc., 1962).

19. Farrar, D. E., The Investment Decision Under Uncertainty, (Englewood Cliffs, N.J., Prentice-Hall Inc., 1962).

20. Fabrychy, W. J. and P. E. Torgersen, Operations Economy, Industrial Applications of Operations Research, (Englewood Cliffs, N.J., Prentice-Hall Inc., 1966).

21. Feldbaum, A. A., Optimal Control Systems, (New York, Academic Press, 1965).

22. Fisher, Irving, The Theory of Interest, (New York, The MacMillan Co., 1930).

23. Gaffney, M. (ed.), Extractive Resources and Taxation, (Madison, Wisconsin, The University of Wisconsin Press, 1968).

24. Gass, S. I., Linear Programming, (New York, McGraw-Hill, Inc., 1958).

25. Gatlin, C., Petroleum Engineering, Drilling and Well Completion, (Englewood Cliffs, N.J., Prentice-Hall Inc., 1960).

26. Gillis, F. E., Managerial Economics: Decision Making Under Certainty for Business and Engineering, (Reading, Mass, Addison-Wessley Press Inc, 1969).

27. Hadley, G., Linear Programming, (Reading, Mass., Addison Wessley Press Inc., 1962).

28. Hadley, G., Nonlinear and Dynamic Programming, (Reading, Mass., Addison-Wessley Press Inc., 1964).

29. Hannsmann, F., Operations Research Techniques and Capital Investment Theory, (New York, John Wiley & Sons Inc., 1969).

30. Hannsmann, F., Operations Research in Production and Inventory Control, (New York, John Wiley & Sons Inc., 1962).

31. Herold, S. C., Analytical Principles of the Production of Oil, Gas, and Water from Wells, (Palo Alto, California, Stanford University Press, 1928).

32. Katz, D., Handbook of Gas Engineering, (New York, McGraw-Hill Inc., 1959).

33. Kaufman, A., Graphs, Dynamic Programming, and Finite Games, (New York, Academic Press, 1967).

34. Kaufman, A. and R. Cruon, Dynamic Programming Sequential Scientific Management, (New York, Academic Press, 1967).

35. Knuth, D. E., The Art of Computer Programming;Volume 1, Fundamental Algorithms; Volume 3, Sorting and Searching; Volume 4, Combinatorial Algorithms; Volume 5, Syntactic Algorithms; Volume 7, Compilers; (Reading, Mass., Addison Wessley Press Inc., 1968-1969).

36. Lewellen, W., The Cost of Capital, (Belmont, California, Wadsworth Press Inc., 1969).

37. Lutz, Friedrick and Vera, The Theory of Investment of the Firm, (Princeton, N.J., Princeton University Press, 1951).

38. Manne, A. S., Scheduling of Petroleum Refinery Operations, (Cambridge, Mass., Harvard University Press, 1956).

39. Manne, A. S. (ed.), Investment for Capacity Expansion - Size, Location, and Time-Phasing, (London, The MacMillan Co., 1967).

40. Marglin, S., The Dynamics of Investment, (Chicago, Rand McNalley Co., 1962).

41. Markowitz, H.M., Portfolio Selection, (New York, John Wiley & Sons Inc., 1959).

42. Muscat, M., Flow of Homogeneous Fluids, (New York, The MacMillan Co., 1949).

43. Muscat, M., Physical Principles of Oil Production, (New York, The MacMillan Co., 1949).

44. Pirson, S. J., Elements of Oil Reservoir Engineering, (New York, McGraw-Hill Inc., 1958).

45. Pirson, S. J., Handbook of Well Log Analysis, Englewood Cliffs, N.J., Prentice-Hall Inc., 1963).

46. Rostow, E. E., A National Policy for the Oil Industry, (New Haven, Conn., Yale University Press, 1948).

47. Scitovsky, T., Welfare and Competition, (Chicago, Richard D. Irwin Inc., 1951).

48. Scott, H. Anthony, Natural Resources, The Economics of Conservation, (Toronto, The University of Toronto Press, 1955).

49. Smith, V. L., Investment and Production, (Cambridge, Mass., Harvard University Press, 1960).

50. Terchroew, D., An Introduction to Management Science, Deterministic Models, (New York, John Wiley & Sons Inc., 1964).

51. Terborgh, G., Dynamic Equipment Policy, (New York, McGraw-Hill, 1949).

52. Tou, J. T., Optimum Design of Digital Control Systems, (New York, Academic Press, 1963).

53. Weingartner, H. Martin, Mathematical Programming and the Analysis of Capital Budgeting Problems, (Chicago, Markham Press, 1967).

54. Wicksell, K., Lectures On Political Economy, (New York, The MacMillan Co., 1934).

55. Zimmerman, E. W., Conservation in the Production of Petroleum, (New Haven, Conn., Yale University Press, 1957).

Journal Articles

1. Agin, N., "Optimum Seeking with Branch and Bound," _Management Science_, (1966), Vol. 13, pp. 176-186.

2. Aronofsky, J. S. and R. Jenkins, "A Simplified Analysis of Unsteady Radical Flow," _Trans. AIME_, (1954), Vol. 201, pp. 147-151.

3. Aronofsky, J. S. and A. Lee, "A Linear Programming Model for Scheduling the Production of Crude Oil," _Trans. AIME_, (1958), Vol. 213, pp. 474-478.

4. Aronofsky, J. S. and A. C. Williams, "The Use of Linear Programming and Mathematical Models in Underground Oil Production," _Management Science_, (1962), Vol. 8, pp. 396-406.

5. Balas, Egar, "An Additive Algorithm for Solving Linear Programs with Zero-One Variables," _Operations Research_, (1965), Vol. 13, pp. 517-546.

6. Balinski, N. L., "Integer Programming: Methods, Uses, Computations," _Management Science_, (1966), Vol. 13, pp. 351-406.

7. Baumol, W. J. and R. Quandt, "Investment and Discount Rates under Capital Rationing -- A Programming Approach," _Economic Journal_, (1965), Vol. 75, pp. 317-329.

8. Bellman, R. E., "Dynamic Programming Solution of the Traveling Salesman Problem," _Journal of the Association for Computing Machines_, (1962), Vol. 9, pp. 61-63.

9. Bernhard, R. H., "Discount Methods for Expenditure Evaluation -- a Clarification of Theory Assumptions," _Journal of Industrial Engineering_, (1962), Vol. 8, pp. 19-27.

10. Broido, A., R. J. McConmen and W. G. O'Regan, "Some Operational Research Applications in the Conservation of Wildland Resources," _Management Science_, (1965), Vol. 11, pp. 802-814.

11. Brownscombe, E. R. and F. Collins, "Pressure Distribution in Unsaturated Oil Reservoirs," _Trans. AIME_, (1950), Vol. 188, pp. 371-374.

12. Burt, G. R., "Optimal Resource Use over Time with an Application to Ground Water," Management Science, (1965), Vol. 11, pp. 80-93.

13. Catchpole, R. A., "The Application of Linear Programming to Integrated Supply Problems in the Oil Industry," Operations Research Quarterly, (1962), Vol. 13, pp. 863-871.

14. Charnes, A., W. W. Cooper and B. Mellon, "Blending Aviation Gasolines," Econometrica, (1952), Vol. 20, pp. 154-168.

15. Charnes, A., W. W. Cooper and M. H. Miller, "Application of Linear Programming to Financial Budgeting and Cost of Funds," Journal of Business, (1959), Vol. 32, pp. 20-46.

16. Chilton, C. H. "Sixteenth Factor Applied to Complete Plant Cost," Chemical Engineering, (1950), pp. 112-114.

17. Cord, J., "A Method for Allocating Funds to Investment Projects when Returns Are Subject to Uncertainty," Management Science, (1964), Vol. 10, pp. 335-341.

18. Cullender, M. A., "The Isochronal Performance Method of Determining the Flow Characteristics of Gas Wells," Trans. AIME, (1955), Vol. 204, pp. 134-138.

19. Dychrnam, T. R., "Allocating Funds to Investment Projects when Returns Are Subject to Uncertainty," Management Science, (1964), Vol. 11, pp. 136-151.

20. Foley, Lyndon L., "Spacing of Oil Wells,"Trans. AIME, (1938), Vol. 127, pp. 15-24.

21. Freund, R., "The Introduction of Risk into a Programming Model," Econometrica, (1956), pp. 258-264.

22. Friedman, M. and L. J. Savage, "The Utility Analysis of Choice Involving Risk," Journal of Political Economy, (1948), Vol. 56, pp. 279-304.

23. Garvin, W., H. Crandall, et. al., "Applications of Linear Programming in the Oil Industry," Management Science, (1957), Vol. 3, pp. 157-168.

24. Gavett, J. and N.Plyter, "The Optimal Assingments of Facilities to Location by Branch and Bound,"

Operations Research, (1966), Vol. 14, pp. 210-232.

25. Golomb, S. W. and L. D. Baumert, "Backtrack Programming," Journal of the Association for Computing Machinery, (1965), Vol. 12, pp. 516-524.

26. Haseman, W. P., "Profits and Proper Spacing of Wells," Oil and Gas Journal, (October 18, 1928), Vol. 27, pp. 52-56.

27. Haseman, W. P., "A Formula Method for Well Spacing and Rate of Production," National Petroleum News, (1929), Vol. 21, pp. 56-59.

28. Haseman, W. P., "A Theory of Well Spacing," Trans. AIME, (1930), Vol. 86, pp. 146-149.

29. Hespos, R. F. and P. A. Strassmann, "Stochastic Decision Trees for the Analysis of Investment Decisions," Management Science, (1965), Vol. 11, pp. 517-530.

30. Hillier, F. S., "The Derivation of Probabilistic Information for the Evaluation of Risky Investments," Management Science, (1963), pp. 443-457.

31. Hirschleifer, J., "On the Theory of Optimal Investment Decision," Journal of Political Economy, (1958), Vol. 66, pp. 329-352.

32. Hitchcock, F. L., "The Distribution of a Product from Several Sources to Numerous Localities," Journal of Mathematics and Physics, (1941), pp. 19-29.

33. Hurst, W., "Unsteady Flow of Fluids in Oil Reservoirs," Physics, (1934), Vol. 5, pp. 19-25.

34. Hwang, C. L., L. T. Fan and L. E. Erickson, "Optimum Production Planning by the Maximum Principle," Management Science, (1967), Vol. 13, pp. 751-755.

35. Ignall, E. and L. Schrage, "Application of the Branch and Bound Technique to Shop Flow-shop Scheduling Problems," Operations Research, (1965), Vol. 13, pp. 400-412.

36. Kaplan, S., "Solution of the Lorie-Savage and Similar Integer Programming Problems by the Generalized Lagrange Multipliers," Operations Research, (1966), Vol. 13, pp. 1130-1136.

37. Kaveler, H. H., "More Wells -- More Oil," The
    Petroleum Engineer, (August-Sept. 1950),
    pp. 135-141.

38. Kolesar, P. J., "A Branch and Bound Algorithm for the
    for the Knapsack Problem," Management Science,
    (1967), Vol. 13, pp. 723-735.

39. Land, A. H. and A. G. Doig, "An Automatic Method of
    Solving Discrete Programming Problems,"
    Econometrica, (1960), Vol. 28, pp. 497-520.

40. Latarie, H. A. "Criteria for Choice Among Risky
    Ventures," Journal of Political Economy, (1959),
    Vol. 67, pp. 114-155.

41. Lomnicki, Z. A. "A Branch and Bound Algorithm for the
    Exact Solution of the Three Machine Scheduling
    Problem," Operations Research Quarterly, (1965),
    Vol. 16, pp. 89-100.

42. Lorie, J. H. and L. J. Savage, "Three Problems in
    Capital Rationing," Journal of Business, (1955),
    Vol. 28, pp. 229-239.

43. MacRoberts, D. J. "Effects of Transient Conditions in
    Gas Reservoirs," Trans. AIME, (1949), Vol. 186,
    pp. 35-39.

44. Manne, A. S., "Capacity Expansion and Probabilistic
    Growth," Econometrica, (1961), Vol. 29, pp. 632-
    649.

45. Markowitz, H. "Portfolio Selection," Journal of
    Finance, (1952), pp. 89-105.

46. McDowell, I., "The Economical Planning Period for
    Engineering Works," Operations Research, (1960),
    Vol. 8, pp. 533-542.

47. Miller, C. C. and A. B. Dynes, "Maximum Reservoir
    Worth -- Proper Well Spacing," Trans. AIME, (1959),
    pp. 334-340.

48. Moyer, Vaughn, "Some Theoretical Aspects of Well
    Drainage and Economic Ultimate Recovery," Trans.
    AIME, (1948), Vol. 174, pp. 88-101.

49. Muscat, M., "The Flow of Compressible Fluids Through
    Porous Media and Some Problems in Conduction,
    Encroachment of Water into an Oil Sand," Physics,
    (1934), Vol. 5, pp. 69-109.

50. Muscat, M., "Principles of Well Spacing," <u>Trans. AIME</u>, (1940), Vol. 136, pp. 131-141.

51. Muscal, M. and M. W. Meres, "The Flow of Heterogeneous Fluids Through Porous Media,"<u>Physics</u>, (1936), Vol. 7, pp. 321-329.

52. Naslund, B.," A Model of Capital Budgeting Under Risk," <u>Journal of Business</u>, (1966), Vol. 39, pp. 257-271.

53. Naslund, B. and A. Whinston, "A Model of Multi-Period Investment Under Uncertainty," <u>Management Science</u>, (1962), Vol. 11, pp. 184-200.

54. Quirk, J. P. and V. L. Smith, "Dynamic Economic Models of Fishing," Research Papers in Theoretical and Applied Economics, University of Kansas, Paper no. 22, (1969).

55. Phelps, Robert W., "Analytical Principles of the Spacing of Oil and Gas Wells," <u>Trans. AIME</u>, (1928-1929), Vol. 82, pp. 90-102.

56. Schilthius, R. J., "Connate Water in Oil and Gas Sands," <u>Trans. AIME</u>, (1938), Vol. 127, pp. 199-214.

57. Sharpe, W. F., "A Simplified Model of Portfolio Selection," <u>Management Science</u>, (1963), Vol. 9, pp. 157-179.

58. Smith, V. L., "Economics of Production from Natural Resources," <u>American Economic Review</u>, (1968), Vol. 57, pp. 409-431.

59. Soloman, E., "Measuring a Company's Cost of Capital," <u>Journal of Business</u>, (1955), pp. 465-495.

60. Teichroew, D., A. A. Robichek and M. Montanbalm, "Mathematical Analysis of the Rate of Return Under Certainty," <u>Management Science</u>, (1965), Vol. 11, pp. 395-403.

61. Theil, H., "Econometrics and Management Science: Their Overlap and Interaction," <u>Management Science</u>, (1965), Vol. 11, pp. 200-212.

62. Thompson, R. G. and M. P. George, "Optimal Operations and Investment of the firm," <u>Management Science</u>, (1966), Vol. 13, pp. B84-B92.

493

63. Van Horne, J., "Capital Budgeting Decisions Involving Combinations of Risky Investments," *Management Science*, (1966), Vol. 13, pp. 58-64.

64. Weingartner, H. Martin, "Capital Budgeting of Interrelated Projects, Survey and Synthesis," *Management Science*, (1966), Vol. 13, pp. 485-516.

65. White, D. J., "Comments on a Paper by McDowell," *Operations Research*, (1961), Vol. 9, pp. 580-584.

66. Wood, D. E., "Branch and Bound Methods," *Operations Research*, (1966), Vol. 14, pp. 699-717.

67. Vergin, C. and J. D. Rogers, "Theoretical and Computational Procedures for Locating Economic Facilities," *Management Science*, (1967), Vol. 13, pp. 240-254.

Other Publications

1. Bartram, J. G., "Well Spacing from a Geologist's Viewpoint," Research and Coordinating Committee, Interstate Oil Compact Commission, (Washington, D. C., 1951).

2. Craze, R. C. and S. E. Buckley, "A Factual Analysis of Effect of Well Spacing on Oil Recovery," API Drilling and Production Practice, (Dallas, Texas, 1946).

3. Craze, R. C. and J. W. Danville, Well Spacing, Humble Oil and Refining Company, Houston, Texas, 1955).

4. Cutler, W. W., Jr., "Estimation of Underground Oil Reserves by Oil-Well Production Curves," U. S. Bureau of Mines, Bulletin No. 228, (Washington, D. C., 1924).

5. Hardwicke, R. E. Antitrust Laws et. al. vs. Unit Operation of Oil or Gas Pools, American Institute of Mining and Metallurgical Engineers, (New York, 1948).

6. Hartung, P. H., "Capital Budgeting with Several Costs of Capital," 13th T.I.M.S. - Conference, (September 1966).

7. Jacoby, N. H. and J. F. Weston, "Factors Influencing Managerial Decisions in Determining Forms of Business Financing," Conference on Research and Business Finance, NBER, (New York, 1952).

8. Joint Progress Report on Reservoir Efficiency and Well Spacing, The Committee on Reservoir Development and Operation of the Standard Oil Company (New Jersey) Affiliated Companies and of the Humble Oil and Refining Company, (New York, 1943).

9. Kaveler, H. H., "Some Considerations in Regulation of Well Spacing," Research and Coordinating Committee, Interstate Oil Compact Commission, (Washington, D. C., 1950).

10. Klinernbey, L. J., "Permeability of Porous Media to Liquids and Gases," API Drilling and Production Practices, (Dallas, 1941).

11. McMurray, W. F. and J. O. Lewis, "Underground Waste in Oil and Gas Fields and Methods of Prevention," U. S. Bureau of Mines, Technical Paper, unnumbered (Washington, D. C., 1916).

12. Miller, H. C. and R. V. Higgins, "Review of Cutler's Rule of Well Spacing and Use of Energy," U. S. Bureau of Mines, Investigation 3439, (Washington, D. C., 1939).

13. More, T. V., R. J. Schilthusis and W. Hurst, "The Determination of Permeability from Field Data," API, Production Decision Proceedings Bulletin, Vol. 211, (Dallas, 1933).

14. Progress Report on Standards of Allocation of Oil Production Within and Among Pools, The Special Committee and Legal Advisory Committee on Well Spacing and Allocation of Production of the Central Committee on Drilling and Production Practices, Division of Petroleum, American Petroleum Institute, (Dallas, Texas, 1942).

15. Shell International Petroleum Co., Inc., The Petroleum Handbook, (London, 1959).

16. Tombinson, W., "Well Spacing and the Use of Energy," Research and Coordinating Committee, Interstate Oil Compact Commission, (Washington, D. C., 1950).

Unpublished Materials

1. Baker, Norman, Linear Programming, Course notes for
    I. E. 535, (Fall, 1968), Lafayette, Indiana.

2. Baker, Norman, Nonlinear and Dynamic Programming,
    Course notes for I. E. 537, (Spring, 1968),
    Lafayette, Indiana.

3. Baker, N. R. and J. S. Yormark, "Resource Allocation,
    Two Dimensional Constraints and Discrete Dynamic
    Programming," Unpublished paper presented at a
    seminar in I. E. 537, (Spring, 1968), at Purdue
    University.

4. Bergendahl., G., "A Combined Linear and Dynamic Program-
    ming Model for Interdependent Road Investment
    Planning," Research Center in Economic Growth,
    Stockhold University, Memorandum no. 58,
    (February, 1968).

5. Hawkins, C. A., Cost of Problem in the Field Price
    Regulation of Natural Gas, Unpublished Ph.D.
    thesis, (Purdue, 1964).

6. Hill, R., Nonlinear and Dynamic Programming, Course
    notes for I. E. 537, (Spring, 1969), Lafayette,
    Indiana.

7. Howard, G. T., Optimal Dynamic Investment, Unpublished
    Ph.D. thesis, (Johns Hopkins, 1967).

8. Smith, K. V., Selection and Revision Decision Rules for
    for Portfolio.Management, Unpublished Ph.D. thesis,
    (Purdue, 1966).

9. Takayama, A., Mathematical Economics, Course notes for
    Economics 660, (Fall, 1968), Lafayette, Indiana.

10. Whinston, A., Mathematical Programming, Course notes
    for Economics 610, (Spring, 1969), Lafayette,
    Indiana.

11. Whinston, A. and C. Van der Panne, Mathematical Pro-
    gramming Techniques, Course notes for Economics
    690G, (Spring, 1968), Lafayette, Indiana.

VITA

NAME: Louis John Joseph Allain

BIRTH: October 12, 1940
                Bath, Maine

CITIZENSHIP: United States of America

EDUCATION: St. Francis Xavier University, 1964
                Economics, B.A., 1964

                Purdue University, 1966
                Economics, M.S., 1966

                Purdue University, 1970
                Economics, Ph.D., 1970

Experience: Purdue University, Teaching Fellow, 1964-1969
                Macroeconomics, Microeconomics, Statistics.

Fellowships: Ford Fellowship, Summer, 1964
                Krannert Fellowship, Summer, 1966
                Krannert Fellowship, Summer, 1968
                David Ross Fellowship, Summer, 1969

PH.D. THESIS: Capital Investment Models of the Oil and Gas
                Industry: A Systems Approach

Present
Occupation: Economist, W. R. Grace & Company,
                New York, New York.

ENERGY IN THE AMERICAN ECONOMY

An Arno Press Collection

Abdallah, Hussein. **The Market Structure of International Oil With Special Reference to the Organization of Petroleum Exporting Countries.** (Doctoral Thesis, University of Wisconsin, 1966). 1979

Allain, Louis John Joseph. **Capital Investment Models of the Oil and Gas Industry.** (Doctoral Thesis, Purdue University, 1970). 1979

Bakerman, Theodore. **Anthracite Coal: A Study in Advanced Industrial Decline.** (Doctoral Dissertation, University of Pennsylvania, 1956). 1979

Barnett, Harold J. **Atomic Energy in the United States Economy.** (Doctoral Thesis, Harvard University, 1952). 1979

Becker, Clarence Frederick. **Solar Radiation Availability on Surfaces in the United States as Affected by Season, Orientation, Latitude, Altitude and Cloudiness.** (Doctoral Thesis, Michigan State University, 1956). 1979

Bentley, Jerome Thomas. **The Effects of Standard Oil's Vertical Integration into Transportation on the Structure and Performance of the American Petroleum Industry, 1872-1884.** (Doctoral Dissertation, University of Pittsburgh, 1974). 1979

Bouhabib, Abdallah Rashid. **The Long-Run Supply of New Reserves of Crude Oil in the United States, 1966-1973.** (Doctoral Dissertation, Vanderbilt University, 1975). 1979

Breed, Alice Gerster. **The Change in Social Welfare from Deregulation: The Case of the Natural Gas Industry.** (Doctoral Dissertation, Boston College, 1973). 1979

Carlson, Rodger D. **The Economics of Geothermal Power in California.** (Doctoral Dissertation, Claremont Graduate School, 1970). 1979

Challa, Krishna. **Investment and Returns in Exploration and the Impact on the Supply of Oil and Natural Gas Reserves.** (Doctoral Dissertation, Massachusetts Institute of Technology, 1974). 1979

Cookenboo, Leslie, Jr. **Crude Oil Pipe Lines and Competition in the Oil Industry** *and* **Costs of Operating Crude Oil Pipe Lines.** 1955/1954

Dale, Alfred George. **Nuclear Power Development in the United States to 1960.** (Doctoral Dissertation, University of Texas, 1975). 1979

Deegan, James F[lournoy]. **An Econometric Model of the Gulf Coast Oil and Gas Exploration Industry.** (Doctoral Dissertation, Southern Methodist University, 1975). 1979

Dillon, Robert John. **Reality and Value Judgment in Policymaking.** (Doctoral Dissertation, University of California, Los Angeles, 1974). 1979

DuBoff, Richard B. **Electric Power in American Manufacturing, 1889-1958.** (Doctoral Dissertation, University of Pennsylvania, 1964). 1979

Eichner, Donald O[scar]. **The Inter-American Nuclear Energy Commission.** (Doctoral Dissertation, The American University, 1969). 1979

Ellis, Theodore John. **Potential Role of Oil Shale in the U.S. Energy Mix.** (Doctoral Dissertation, Colorado State University, 1972). 1979

Erickson, Edward Walter. **Economic Incentives, Industrial Structure and the Supply of Crude Oil Discoveries in the U.S., 1946-58/59.** (Doctoral Dissertation, Vanderbilt University, 1968). 1979

Fenichel, Allen Howard. **Quantitative Analysis of the Growth and Diffusion of Steam Power in Manufacturing in the United States, 1838-1919.** (Doctoral Dissertation, University of Pennsylvania, 1964). 1979

Foster, Abram John. **The Coming of the Electrical Age to the United States.** (Doctoral Dissertation, University of Pittsburgh, 1952). 1979

Fulda, Michael. **Oil and International Relations.** (Doctoral Dissertation, The American University, 1970). 1979

Gessford, John Evans. **The Use of Reservoir Water for Hydroelectric Power Generation.** (Doctoral Dissertation, Stanford University, 1957). 1979

Gilbreth, Terry John. **Governing Geothermal Steam.** (Doctoral Dissertation, University of California, Riverside, 1974). 1979

Grayson, C. Jackson, Jr. **Decisions under Uncertainty: Drilling Decisions by Oil and Gas Operators.** 1960

Hall, Harry S. **Congressional Attitudes Toward Science and Scientists.** (Doctoral Dissertation, University of Chicago, 1961). 1979

Jacoby, Henry Donnan. **Analysis of Investment in Electric Power.** (Doctoral Thesis, Harvard University, 1967). 1979

Johnson, Charles J. **Coal Demand in the Electric Utility Industry, 1946-1990.** (Doctoral Thesis, Pennsylvania State University, 1972). 1979

Johnson, James P. **A "New Deal" for Soft Coal.** (Doctoral Dissertation, Columbia University, 1968). 1979

Keating, William Thomas. **Politics, Technology, and the Environment.** (Doctoral Dissertation, Indiana University, 1974). 1979

Kolb, Jeffrey Alan. **An Econometric Study of Discoveries of Natural Gas and Oil Reserves in the United States, 1948 to 1970.** (Doctoral Dissertation, University of Oregon, 1974). 1979

Lawrence, Anthony G. **Pricing and Planning in the U.S. Natural Gas Industry.** (Doctoral Dissertation, State University of New York at Buffalo, 1973). 1979

Lehman, Edward Richard. **Profits, Profitability, and the Oil Industry.** (Doctoral Dissertation, New York University, 1964). 1979

Manes, Rene Pierre. **The Effects of United States Oil Import Policy on the Petroleum Industry.** (Doctoral Thesis, Purdue University, 1961). 1979

Marcus, Kenneth Karl. **The National Government and the Natural Gas Industry, 1946-56.** (Doctoral Thesis, University of Illinois, 1962). 1979

McDonald, Philip R. **Factors Influencing Fuel Oil Growth.** (Doctoral Thesis, Harvard University, 1966). 1979

Meloe, Torleif. **United States Control of Petroleum Imports.** (Doctoral Dissertation, Columbia University, 1966). 1979

Nowill, Paul Henry. **Productivity and Technological Change in Electric Power Generating Plants.** (Doctoral Dissertation, University of Massachusetts, 1971). 1979

Pagoulatos, Angelos. **Major Determinants Affecting the Demand and Supply of Energy Resources.** (Doctoral Dissertation, Iowa State University, 1975). 1979

Pendergrass, Bonnie Baack. **Public Power, Politics, and Technology in the Eisenhower and Kennedy Years.** (Doctoral Dissertation, University of Washington, 1974). 1979

Phillips, David Gordon. **Federal-State Relations and the Control of Atomic Energy.** (Doctoral Dissertation, Syracuse University, 1964). 1979

Schramm, Gunter. **The Role of Low-Cost Power in Economic Development.** (Doctoral Dissertation, University of Michigan, 1967). 1979

Simon, Simon M. **Economic Legislation of Taxation.** (Doctoral Thesis, New York University, 1968). 1979

Smith, David Brian. **The Economics of Inter-Energy Competition in the United States.** (Doctoral Dissertation, University of Nebraska, 1971). 1979

Spann, Robert M. **The Supply of Natural Resources.** (Doctoral Thesis, North Carolina State University at Raleigh, 1970). 1979

Spooner, Robert Donald. **Response of Natural Gas and Crude Oil Exploration and Discovery to Economic Incentives.** (Doctoral Dissertation, University of Pennsylvania, 1973). 1979

Steele, Henry. **The Economic Potentialities of Synthetic Liquid Fuels from Oil Shale.** (Doctoral Dissertation, Massachusetts Institute of Technology, 1957). 1979

Striner, Herbert E. **An Analysis of the Bituminous Coal Industry in Terms of Total Energy Supply and a Synthetic Oil Program.** (Doctoral Dissertation, Syracuse University, 1951). 1979

Strout, Alan Mayne. **Technological Change and United States Energy Consumption, 1939-1954.** (Doctoral Dissertation, University of Chicago, 1967). 1979

Waltrip, John Richard. **Public Power During the Truman Administration.** (Doctoral Dissertation, University of Missouri, 1965). 1979

Wedemeyer, Karl Eric. **Interstate Natural Gas Supply and Intrastate Market Behavior.** (Doctoral Dissertation, University of Southern California, 1972). 1979

Whillier, Austin. **Solar Energy Collection and its Utilization for House Heating.** (Doctoral Thesis, Massachusetts Institute of Technology, 1953). 1979

Young, James Van. **Judges and Science: The Case Law on Atomic Energy.** (Doctoral Dissertation, University of Iowa, 1964). 1979